Biofuels

Related Titles

Biopolymers Based Advanced Materials
ISBN: 978-0-6482205-4-1 (e-book)
ISBN: 978-0-6482205-5-8 (hardcover)

Functional Polymer Blends and Nanocomposites
ISBN: 978-0-6482205-6-5 (e-book)
ISBN: 978-0-6482205-7-2 (hardcover)

Polymers in Oil and Gas Industry
ISBN: 978-0-6482205-0-3 (e-book)
ISBN: 978-0-6482205-1-0 (softcover)

Functional Nanomaterials and Nanotechnologies: Applications for Energy & Environment
ISBN: 978-0-6482205-2-7 (e-book)
ISBN: 978-0-6482205-3-4 (softcover)

Technology Management in Business
ISBN: 978-1-925823-02-8 (softcover)

Advances in Polymer Technology: Material Development, Properties and Performance Evaluation
ISBN: 978-1-925823-00-4 (e-book)
ISBN: 978-1-925823-01-1 (hardcover)

Polymer Nanomaterials for Specialty Applications
ISBN: 978-1-925823-03-5 (e-book)
ISBN: 978-1-925823-04-2 (hardcover)

Advanced Materials
ISBN: 978-1-925823-05-9 (e-book)
ISBN: 978-1-925823-06-6 (hardcover)

Engineering Chemistry
ISBN: 978-1-925823-10-3 (softcover)

Biofuels

Dr. Vikas Mittal
Editor

CWP

Central West Publishing

NATIONAL LIBRARY OF AUSTRALIA

A catalogue record for this book is available from the National Library of Australia

ISBN (print): 978-1-925823-13-4
ISBN (e-book): 978-1-925823-12-7

Contents

Preface

Increased global warming, high consumption of fossil fuels, environmental pollution and decreasing petroleum reserves necessitate the need of energy production from environmentally friendly sources for fulfilling the demands of the industrialized world. In this respect, various alternative energy sources have been developed in the recent years, including a family of biofuels. Many biofuels exhibit similar properties as compared to their fossil-based counterparts and can be very beneficial in the future for replacing these fossil fuels. As a result, the field of biofuels has seen major breakthroughs in the recent years with respect to synthesis, properties and applications of the high-quality biofuels. Though many challenges exist till date which hinder the large scale commercialization of these materials, however, it is envisaged that growth in the technological knowhow as well as new feedstocks in the near future will continue to drive the development of biofuels. The book aims to present the recent developments in various aspects of such biofuels, especially related to feedstocks, synthetic methodologies as well as commercialization.

Chapter 1 reviews the classification and generations of biofuels, along with presenting feedstocks and techniques for the development of biofuels. In Chapter 2, a brief review on the catalytic upgrading of furfural and levulinic acid for the production of alternative fuels has been presented. Chapter 3 highlights the recent advances and future opportunities for producing low-cost renewable biofuels at industrial scale by expression of biosynthetic pathways in bacteria, yeast, algae and cyanobacteria. Chapter 4 summarizes the supercritical water gasification process of biomass. Several aspects for technology upgrading such as reactor configurations and operating conditions (temperature, pressure, feed concentration and catalyst), which have crucial role in maximizing H_2 production as well as gasification efficiency, have been discussed. Chapter 5 investigates the production of biodiesel from green microalgae applying hexane as an extracting solvent via Soxhlet extractor. Effect of inorganic fertilizer, employed as nutrient treatment during culturing of the freshwater green microalgae, was also studied. Chapter 6 takes a look at some of the technologies that are at the forefront of what might be called an "agro-industrial revolution" in its true sense. It is concluded that research efforts are now beginning to bear fruit, as a steady growth is seen in the number and scale of biorefineries capable of economically and sustainably converting lignocellulosic biomass into bioethanol.

The editor is indebted to the chapter authors for providing their deep insights into various aspects of biofuels. The book would not have been successfully accomplished without their support. The book is dedicated to my family for constant motivation and encouragement.

Vikas MITTAL

Chapter 1

Biofuels: Classification, Generations and Feedstocks/ Techniques for Biofuels Generation

Naman Arora and Vikas Mittal*,**
Department of Chemical Engineering, The Petroleum Institute (a part of Khalifa University of Science and Technology), Abu Dhabi, UAE
Current address: Bletchington, Wellington County, Australia
**Corresponding author*: vik.mittal@gmail.com

1.1 Introduction

Numerous factors, such as increased global warming, high consumption of fossil fuels, environmental pollution and decreasing petroleum reserves, necessitate the need of energy production from environmentally friendly sources for fulfilling the demands of the industrialized world. Some alternative sources of energy like wind, solar, biofuels, etc., have exhibited promising results to replace fossil fuels. Biofuel like biodiesel is a promising replacement of petroleum fuel since it is non-toxic, renewable, biodegradable and emits less harmful gases like sulfur oxide [1-4]. Transesterification is the most employed technique for the production of biodiesel which involves a vegetable oil or animal fat, methanol or any other alcohol (ethanol and butanol) and a homogeneous or heterogeneous catalyst. Biodiesel can be utilized in the existing diesel engines without any modification requirement as the biofuels and fossil fuels have almost similar properties. Currently, many European countries and United States employ biodiesel and bioethanol in different applications. Other nations like Malaysia also utilize the blend consisting of 95% petroleum diesel and 5% palm oil. In future, it is envisaged that the consumption of biodiesel will grow as an alternative to fossil fuels [5-9]. Higher cost of processing of biodiesel is still the main limitation for its extensive commercialization. There are numerous factors which decide the cost of biodiesel including storage terminal, purification process, raw materials, reactants, etc. Among these components, feedstock alone costs 80% of the complete processing cost of biodiesel. Therefore, it is viable to achieve a low-cost biodiesel by adapting to economical feedstocks like waste fats and oils [3,10]. Additionally, there are other challenges associated with the usage of biofuels, which need to be overcome to realize large scale usage of these materials [11].

1.2 Classification of Biofuels

Biofuels are generally categorized as primary and secondary fuels. The sources of primary biofuels are wood chips, fuel-wood and pellets which are employed in an unprocessed form for the production of electricity as well as for cooking and heating purposes. Thus, primary biofuels are naturally existing unprocessed biomass which is consumed without making any modification in its natural form. On the other hand, biomass processing is performed for generating secondary biofuels. This type of biofuel can be obtained in the liquid, solid or gas form such as charcoal, biogas, bio-oil, bioethanol, hydrogen and biodiesel. Secondary biofuels are utilized in numerous industrial applications, transportation and other processing purposes. Secondary biofuels are further classified as first-, second- and third-generation biofuels based on the type of raw materials and production methodologies [12,13].

1.2.1 First-generation Biofuels

In this category, grains [14-17], sugars [18-24] and seeds are the primary sources for biofuel production. This type of biofuels involves simple processing for obtaining the final product. Ethanol is one of the most common examples of first-generation biofuels which is produced by the fermentation process of sugar derived from different plants and crops containing starch. Another popular example of first-generation biofuels is biodiesel which is prepared from vegetable oils of oleaginous plants by employing transesterification method. In various countries, first-generation biofuels are generated in remarkable commercial amounts. The feasibility of producing first-generation biofuels is, however, challenging due to strong contradiction with the supply of food [13]. Operating cost of producing these biofuels is also high owing to the competition with food demand associated with primary crops like oil-seeds, grains and sugars. These challenges signify the need for identifying the non-edible sources of biomass for obtaining first-generation biofuels, which can also help to maintain sustainable supply of food for the communities, especially in the poorer regions and countries.

1.2.2 Second-generation Biofuels

Thermochemical and biological processing based techniques usually produce the second-generation biofuels. These type of biofuels are extracted from non-eatable plant biomass and non-edible residual matter of different crops. Basically, specific biomass is generated for producing second-generation biofuels to fulfill the demand of energy which does not involve the edible crops. Classification of second-generation biofuels can be made on the basis of biochemical and thermochemical processes which are employed for converting biomass into biofuels. Based on the biochemical process, biofuels like bioethanol, biodiesel and butanol are generated. On the other hand, fuels like dimethyl ether, Fischer-Tropsch liquids (FTL) and methanol are generated by using the thermochemical process.

Larson [12] stated that the basic properties of feedstocks can lower the processing cost and have potential energy and environmental advantages for most of the second-generation biofuels. Thegarid *et al.* [25] reported the generation of second-generation biofuels where bio-oil was produced by using catalytic pyrolysis in catalytic cracking co-processing mode with vacuum gas oil. The study confirmed the possibility for the development of biofuels through co-processing in fluid catalytic cracking unit. Schenk *et al.* [26] reported microalgae based second-generation biofuels. The benefits of second-generation microalgal systems are that they generate a broad variety of feedstocks for producing bio-methane, biodiesel, bio-hydrogen and bioethanol. The important benefits associated with microalgal system are: a) it exhibits a higher photon conversion capability, b) this type of system can be harvested batch-wise almost throughout a year which imparts constant oil supply, c) this system is environment-friendly as it only requires limited amount of freshwater and only consumes waste water and salt water, d) it is capable of coupling carbon-dioxide-neutral fuel production with carbon-dioxide sequestration and e) this system generates biodegradable and toxicity-free biofuels. Process of harvesting and delivery of carbon-dioxide are the main limitations for the potential production of the biofuels. Dias *et al.* [27] analyzed the simulation results associated with different stages of coupling between first- and second-generation ethanol based setups using modified hydrolysis technology and pentoses fermentation. It was revealed that the merged first- and second-generation exhibited numerous benefits over independent and unmerged second-generation ethanol production setups.

1.2.3 Third-generation Biofuels

As discussed earlier, pursuing first-generation biofuels may result in immense pressure on the commercial food markets all around the world. Moreover, fuels associated with first-generation lead to the deterioration of the plant and water shortages. Some of the challenges associated with the first-generation biofuels were effectively resolved by the introduction of second-generation biofuels. However, the challenges of land usage and associated changes [28] still limit the application of biofuels associated with second-generation. Therefore, in order to overcome the limitations of first- and second-generation biofuels, a latest approach (third-generation biofuels) is aimed towards using microscopic organisms as feedstock instead of agricultural residues and vegetable oils. Biofuels associated with third-generation are extracted from microbes and microalgae which have proved to be a reliable source of energy to replace petroleum-derived fuels.

The consolidated bioprocessing (CBP) system is capable in reducing the cost of biofuels because of its simple feedstock processing and lower energy input needs. In this system, the production of cellulose, fermentation and hydrolysis of substrate are carried out in one process step through cellulolytic microorganism. Therefore, CBP system is an economical approach towards the production of third-generation biofuels [29]. Saidane-Bchir *et al.* [30] reported the production of third-generation bioethanol by using the isolated strain of microalgae which was extracted from the wastewater of a slaughterhouse. It was revealed that the strain of microalgae is a potential feedstock for producing third-generation bioethanol. Zhu *et al.* [31] reported the potential production of lipid content by using Trichosporon fermentans and subsequently, it was used for producing biodiesel. It was observed that in pre-treated waste molasses, Trichosporon fermentans evolved or developed effectively with a yield of obtained lipid content of 12.8 g/L in the molasses of 15% total sugar concentration. The study demonstrated that yeast is a promising source for the preparation of inexpensive microbial oil from the residues of agriculture based industries.

Samori *et al.* [32] reported the extraction of hydrocarbons from Botryococcus braunii by employing switchable-polarity solvents. The system of switchable solvent comprised of 1,8-diazabicyclo-[5.4.0]-undec-7-ene (DBU) and an alcohol. The authors examined two different systems of switchable-polarity solvents, DBU-octanol and DBU-ethanol, for extracting hydrocarbons from the samples of Botryococcus braunii. The system based on DBU-octanol exhibited better extraction efficiency as compared to conventional solvents like methanol, n-hexane, etc. Thus, the usage of switchable-polarity solvents represents a beneficial route for harvesting the algal biomass as it is environmentally friendly and lowers the overall cost of biodiesel production. El-Shimi *et al.* [33] developed biodiesel from Spirulina-Platensis microalgae by employing the acid catalyzed in-situ transesterification technique. The authors studied factors such as amount of methanol, acid catalyst concentration, stirring speed, temperature and time, which could immensely affect the yield and cost of biodiesel production.

Chng *et al.* [34] successfully produced bioethanol from lipid-extracted biomass of Scenedesmus dimorphus. In this work, biomass was directly introduced to simultaneous saccharification and fermentation, thereby, avoiding the expensive pre-treatment, reducing the possibility of contamination and decreasing the complexity of high sugar content. Abreu *et al.* [35] reported the strategy to decrease the cost of producing microalagal biomass, thus, developing cheaper biodiesel fuel. In this study, chlorella vulgaris was grown under both mixotrophic and photoautotrophic conditions. In order to grow chlorella vulgaris, cheese whey was utilized as a source of carbon. It was observed that mixotrophic microalgae evolved more rapidly as compared to photoautotrophic cells. Usage of industrial dairy waste (cheese whey) also contributed towards overcoming the environmental issues associated with the disposal of cheese whey in the diary industries.

In a recent study, Srinophakun *et al.* [36] reported the strategy to produce feedstock for the preparation of biofuels by cultivating Arthrobacter Ak19 in the presence of stress agent like potassium permanganate. Herein, glycerol was used as a source of pure carbon. Response surface design was employed for the optimization of the proportion of nutrients for enhancing the concentration of biomass. It was observed that with the usage of potassium permanganate (stress agent) at a concentration of 3.2 mg/L, 1.95 times higher content of lipid was obtained as compared to the lipid content achieved under normal conditions. In another study, Wang *et al.* [37] also reported the preparation of biodiesel from corn oil by using lipozyme TL IM catalyzed methanolysis. Similar to the other study, the authors reported that the production of biodiesel improved by employing response surface methodology.

1.3 Types of Liquid Biofuels

1.3.1 Biodiesel

Biodiesel is extracted from organic feedstock like waste animal fats or vegetable oils. On burning, biodiesel emits lower percentage of harmful gases as compared to petroleum-derived diesel. The usage of biodiesel does not directly or indirectly raise the level of carbon dioxide in the environment, thereby, contributing to the reduction of the greenhouse effect. Furthermore, biodiesel exhibits better properties as compared to petroleum-derived diesel in respect to the content of sulfur, aromatic content and flash point [38,39]. However, biodiesel cannot currently entirely replace the petroleum-derived diesel, though its importance is steadily increasing as reserves of petroleum are reducing with time. Both edible and non-edible vegetable oils have been employed to produce biodiesel. Some of the common examples of edible vegetable oils are corn and canola, whereas non-edible vegetable oils such as madhuca indica and Pongamia pinnata are commonly employed [40-42]. As biodiesel is usually seen as an effective alternative for petroleum-based diesel fuel, thus, it is critical to examine its key features discussed below.

Chemical structure: Petroleum-extracted diesel is comprised of combinations of a wide variety of hydrocarbons (C12-C25) made up of aromatics, napthenes and paraffins. Biodiesel is a mixture of a small variety of molecules which are esters of fatty acids of C12, C14, C16, C18 and C22. Ring structures like aromatic molecules are exhibited by diesel fuel, whereas straight chain hydrocarbon esters constitute the primary structure of a variety of biodiesels.

Antifoaming: It is convenient to fill pure biodiesel into the fuel tanks of vehicles as it exhibits superior anti-foaming property. As a consequence, foam generation is minimized, which avoids overflowing and leakage of the biodiesel out of the fuel tank. On the other hand, petroleum-extracted diesel has poor anti-foaming property.

Oxygen content: Petroleum-derived diesel does not hold any oxygen content. On the other hand, biodiesel usually exhibits 11% of oxygen content and exhibits properties such as conductivity, solvency and detergency. Due to this reason, biodiesel can also create linkages with metal surfaces (wettability), as it is polar in nature.

Cetane number: The cetane number of petroleum-extracted diesel and biodiesel is usually observed to be in the range of 40-52 and 45-70 respectively. Distribution pattern of fatty acids present in oils and fats decides the cetane number of biodiesel. High value of cetane number can be achieved by fatty acids which are longer and more saturated.

Conductivity: Biodiesel exhibits lower risk of fires and sparks induced by static charge as pure biodiesel has high conductivity, which exceeds 500 picoS/m.

Corrosion: Generally, corrosion occurs due to the presence of oxygen which is associated with the absorption of water. However, the wettability property of pure biodiesel opposes the corrosion process by decreasing the movement of oxygen molecules to the metal surface. Additionally, pure biodiesel does not contain any sulphur content and corrosive entities, which subsequently avoids the corrosion of copper and yellow metals.

Cold flow properties: In pure biodiesel, solidification process is very fast as biodiesel is a simple mixture of low number of constituents. On the other hand, the solidification process of petroleum-derived diesel is relatively slow as the crystallization temperatures of the constituents present in diesel have different values [43].

1.3.2 Bioethanol

Bioethanol results due to the utilization of renewable biomass for producing ethanol. Bioethanol is harmless, environmental-friendly and renewable fuel [44]. In 2006, 51 billion litres of bioethanol based on first-generation processes were produced globally. India and China produced 11% of the global generation of bioethanol in 2006. Recently, most of the countries have focused on the development of first-generation bioethanol. Blending of bioethanol and petrol has also been observed to exhibit complete burning of fuel inside the existing transportation vehicles. A litre of ethanol generates only 66% of energy produced by the same amount of petrol, however, ethanol exhibits higher octane number. Moreover, it demonstrates better performance when blended with petroleum-extracted fuel as compared to pure petrol. The combustion process of the fuel in engines has been observed to improve with the usage of ethanol due to the reduction in partially burned hydrocarbons and carbon monoxide emissions. An additional benefit of utilizing ethanol-petrol blends is decreased sulphur oxide emissions [45,46].

One of the simplest methods for producing bioethanol is the direct fermentation of biomass containing sugars into ethanol [47,48]. Sindhu *et al.* [49] reported the generation of bioethanol from bamboo (Dendrocalamus sp.) after dilute acid pre-treatment and enzymatic saccharification. In another study, Trivedi *et al.* [50] reported the generation of bioethanol from green seaweed Ulva fasciata. Kwon *et al.* [51] used spent coffee grounds as feedstock for the co-generation of bioethanol and biodiesel. In another recent study, Enquist-Newman *et al.* [52] proposed the potential approach for producing bioethanol from brown macroalgae sugars. Kumar *et al.* [53] reported the utilization of Gracilaria verrucosa as a feedstock for the generation of bioethanol. Borines *et al.* [54] used the macroalgae Sargassum *spp* as a potential renewable feedstock for the purpose of generating bioethanol. Shukla *et al.* [55] also developed the process for producing bioethanol by using waste algal biomass of Gracilaria verrucosa as a feedstock. Raghavi *et al.* [56] developed the sequential approach for generating bioethanol by utilizing sugarcane trash. In another study, Klein *et al.* [57] also demonstrated the microwave irradiation facilitated process for generating bioethanol by using the leaves of Ficus religiosa as feedstock.

1.3.3 Butanol

Butanol is 4-carbon alcohol ($C_4H_{10}O$) based biofuel [58]. It can be effectively blended with gasoline and other hydrocarbons and exhibits higher amount of heat energy as compared to ethanol, equivalent to 25% enhanced harvestable energy (BTU). Butanol has 110,000 BTUs per gallon which is almost equal to the BTUs of gasoline (115,000) [59]. Butanol is less corrosive than bioethanol, thereby, making it more likely to be transported or distributed by using existing pipelines and fuel stations. Butanol is less likely to evaporate as compared to ethanol and gasoline which makes it more secure. Acceptability of butanol-gasoline blend in conventional engines and generation of less volatile organic compound emissions are some of the benefits associated with the usage of butanol-gasoline blends [60]. The burning of butanol is cleaner than ethanol as it contains 22% oxygen [61]. Butanol is commonly observed to exist in the form of four isomers, i.e., n-butanol, 2-butanol, t-butanol and i-butanol referred to as normal-butanol, secondary-butanol, ter-butanol and iso-butanol, respectively. Same energy is exhibited by all isomers, however, different manufacturing techniques are used for these isomers [58]. Jiang *et al.* [62] studied the production of n-butanol from sugarcane juice by utilizing Clostridium acetobutylicum JB200, which is a mutant with high butanol tolerance and potential to produce high-titer n-butanol from glucose. Moradi *et al.* [63] applied alkali and acid pre-treatments for the improved generation of butanol from rice straw. Herein, Clostridium acetobutylicum bacterium was utilized to hydrolyze and ferment rice straw into acetone, butanol and ethanol. In another study, Du *et al.* [64] reported the generation of butanol from acid hydrolyzed corn fiber with Clostridium beijerinckii. Xin *et al.* [65] also developed the process of fermentation for generating butyl-butyrate and butanol by utilizing Clostridium sp. strain. Kong *et al.* [66] used gamma-valerolactone for hydrolyzing the sugarcane bagasse to generate bio-butanol. In another study, Tan *et al.* [67] reported the generation of butanol, ethanol and acetone from barley based dried distillers grains with solubles (DDGS).

1.4 Processes Involved for the Generation of Biodiesel

In order to generate biodiesel, transesterification reaction takes place between fats or oils and alcohol (ethanol or methanol). Generally, two methods are used for the transesterification process, i.e., catalytic transesterification and non-catalytic transesterification. The presence of catalyst is significant as alcohol has very low solubility in oil or fat. Thus, the main function of the catalyst is to improve alcohol solubility, thereby, enhancing the rate of reaction. Catalytic transesterification method involves base catalytic transesterification and acid catalytic transesterification techniques. Selecting the proper catalyst is a crucial factor for lowering the cost of biodiesel generation [68]. Quantity of free fatty acid (FFA) present in the oil is another factor which decides the selection of catalyst type employed for the reaction. In the case of low amount of FFAs present in the oil, base catalyzed esterification is suitable, whereas oils with high amount of FFAs must be processed by acid catalytic transesterification method [69]. Some examples of base catalysts are sodium hydroxide and potassium hydroxide, whereas acid catalysts include sulphuric acid and phosphoric acid. Supercritical alcohol transesterification and BIOX co-solvent transesterification are the methods associated with non-catalytic transesterification process.

Conventionally, homogeneous acid or base catalysts are employed for the process of transesterification to produce biodiesel. Heterogeneous catalytic transesterification is attaining much attention recently as the heterogeneous catalyst does not lead to the formation of soap [70]. Temperature also strongly influences the rate of transesterification reaction. As temperature increases, the rate of reaction also increases and the time of reaction decreases. Solid base and acid catalysts are utilized in heterogeneous catalytic transesterification process, where solid acid catalysts are considered to be less active as compared to solid base catalysts. Hydrotalcites, zeolites and alkaline earth metal oxides are some examples of solid base catalysts, whereas solid acid catalysts include tungstated zirconia, sulfated zirconia and Nafion-NR50. Solid base catalysts ensure reasonable cost for producing biodiesel and utilization of these catalysts in the form of fixed bed reactor set-up leads to the catalyst separation from the product, which have been transesterified [71]. In spite of exhibiting lower activity, solid acid catalysts are employed in various processes associated with industry due to the following benefits: a) no influence of the content of FFA, b) simultaneous esterification and transesterification [72], c) removal of purification stage while generating biodiesel [73], d) effective separation of catalyst from the end products and e) corrosion control [74]. Various catalytic and non-catalytic transesterification processes employed for the generation of biodiesel are explained below.

1.4.1 Alkali Catalytic Transesterification

The base catalytic transesterification process is usually quick, however, it can be easily influenced in contact with the free fatty acids (FFAs) present in oil or fat and water. Reactions of base catalysts with FFAs lead to the formation of soap and water. Formation of soap lowers the generation of alkyl esters and creates complications for separating glycerol and biodiesel as well as water washing as emulsion is formed. It has been noticed that methoxide catalysts exhibit high productivity as compared to hydroxide catalysts. Potassium based catalysts are also observed to be better than sodium based catalysts in terms of producing biodiesel [75].

1.4.2 Acid Catalytic Transesterification

The systems based on acid catalysts have a slower rate of reaction. In acid catalyzed reactions, free fatty acids are converted to esters. Soaps can also be transformed to esters as a pre-treatment step for feedstocks containing high content of free fatty acids [76]. Excess quantity of alcohol is required by the acid catalytic transesterification process, therefore, bigger distillation column and reactor are required as compared to alkali catalyzed process for generating same quantity of biodiesel. The acid catalyzed transesterification provides high conversion efficiency through the increased molar ratio of alcohol to oil, acid catalyst concentration, time of reaction and reaction temperature [77].

1.4.3 Biocatalytic Transesterification

Lipases are the most common examples of naturally available biocatalysts which have the potential for performing the transesterification reactions required for generating biodiesel. These lipases can be extracted from the various species of bacteria, i.e., Rhizomucor miehei, Candida Antarctica, Candida rugosa, etc. [78]. Enzymatic biocatalysts can be classified into two major categories: a) extracellular lipases and b) intracellular lipases. In case of extracellular lipases, enzymes are extracted from the microorganism broth. Mucor miehei, P. cepacia and R. oryzae are the common microorganisms utilized for extracting enzymes. Intracellular lipases are observed to remain inside the cell-producing walls, where enzyme is immobilized. Various methods are involved for the immobilization of lipase such as encapsulation, covalent bonding, entrapment, cross-linking and adsorption. The purpose of employing these methods is to enhance the stability of lipase for generating biodiesel. Among these methods, adsorption is widely utilized for the immobilization process [79]. Despite the high specificity and purity of enzymatic reactions, these reactions are very slow requiring a reaction time varying from 4 to 40 h at temperatures ranging between 35 and 45 °C.

Overall, biocatalytic transesterification process is promising for the generation of biodiesel, however, overcoming its high operational cost is necessary to make it viable for large-scale commercial production.

1.4.4 Non-catalytic Techniques

The catalytic processes for the generation of biodiesel involve various steps associated with ester purification and separation of unreacted reactants and catalysts. As an alternative, biodiesel generation from triglycerides by employing non-catalytic reactions also represents a simple approach. Supercritical alcohol and BIOX processes are two main non-catalytic transesterification methods for generating biodiesel.

Supercritical Alcohol Transesterification Process

Supercritical alcohol process is one of the non-catalytic methods employed for generating biodiesel in which temperature and high pressure are the main factors employed for carrying out the transesterification reaction in place of catalysts [80]. The triglycerides transesterification by employing supercritical ethanol and other alcohols (like methanol, butanol and propanol) has been observed to demonstrate effective output. The reaction is observed to be fast and the conversion efficiency increases from 50% to 95% for the initial 10 minutes with the temperature ranging between 250 and 400 °C. Supercritical methanol transesterification process has been designed in such a way that it can overcome the lag time of initial reaction which occurs due to the lower solubility of alcohol in triglyceride. Separate phases do not exist when the solvent is exposed to temperature and pressure exceeding the critical point.

Supercritical alcohol process is observed to be beneficial as there is no presence of catalyst, which makes the recovery of glycerol and purification of biodiesel relatively easier and environment-friendly [81].

BIOX Co-solvent Transesterification Process

BIOX co-solvent process has gained significant attention for generating biodiesel. In this process, both triglycerides and free fatty acids are transformed in a two-step, single phase and continuous process at atmospheric pressures and near ambient temperatures [43]. The poor solubility of alcohol in triglyceride causes slow rate of reaction which can be overcome by employing co-solvent process. Biox co-solvent process exhibit benefits as it utilizes inert and retrievable co-solvents in a single pass and fast reaction at ambient temperature and pressure. Similar to the supercritical alcohol process, the catalyst does not emerge in biodiesel as well as glycerol [82]. Almost all oils or fats such as crude vegetable oils, animal fats and waste cooking oils can be processed by using this technique.

1.5 Different Feedstocks and Techniques for Producing Biofuels

Zheng *et al.* [83] reported the preparation of biodiesel from unusable frying oil. In this study, the authors investigated the reaction kinetics of the acid catalyzed transesterification of oil in the presence of methanol for generating fatty acid methyl esters. It was observed that reaction kinetics followed pseudo first-order model. In another study, Demirbas and Demirbas [84] reported the significance of algae oil for potential utilization as a source for producing biodiesel. Algae are observed to grow faster than other plants like terrestrial crops and almost half of the weight of algae is oil. In addition, algae is easily available almost everywhere, for example in waste water, salt water, around the rivers or lakes, sewage, etc. Thus, no agricultural land and crops are required in order to obtain algae for biodiesel generation. Moreover, algae requires less energy for processing as compared to the energy provided by it. Overall, algae based biofuels are envisaged to be a better alternative to petroleum-derived fuels in the near future.

Leung and Guo [85] carried out the alkaline catalyzed transesterification of canola oil and waste frying oil, along with studying the properties of three catalysts (sodium methoxide, potassium hydroxide and sodium hydroxide). Content of ester, acid value and kinematic viscosity were the main factors, which controlled the properties of the biofuel generated using three catalysts. It was observed that sodium hydroxide was most beneficial among these catalysts due to its lower price and intermediate catalytic response. Another study was reported by Rashid and Anwar [86] for producing biodiesel by alkaline catalyzed transesterification of rapeseed oil. ASTM (American Society for Testing and Materials) and European EN standards confirmed that the as-prepared biodiesel from rapeseed oil exhibited the prescribed fuel properties.

Dube *et al.* [87] employed the two-phase membrane reactor for the production of biodiesel or fatty acid methyl esters from methanol and canola oil. The main aim of using the two-phase membrane reactor was to eliminate the unreacted canola oil from fatty acid methyl esters (FAME), thereby, producing the biodiesel of pure quality and shifting the reaction equilibrium to the product side. Berchmans and Hirata [88] developed the method for producing biodiesel from crude Jatropha curcas seed oil containing excessive percentage of free fatty acids. In this study, two stage transesterification was applied for improving the production of methyl ester. The first stage was associated with the usage of acid, which decreased the level of free fatty acids of Jatropha curcas seed oil to less than 1%. The second stage was transesterification with alkaline catalyst which provided 90% yield of methyl esters of fatty acids. In another study, Kouzu *et al.* [89] conducted the transesterification of soybean oil with methanol in the presence of solid base catalyst for the purpose of producing biodiesel. Selected solid base catalysts were calcium oxide, calcium hydroxide and calcium carbonate. The yield of fatty acid methyl ester was recorded as 93%, 12% and 0% respectively after 1 hour reaction time.

Liu *et al.* [90] conducted the transesterification of soybean oil with methanol in the presence of solid base catalyst (calcium methoxide) for the production of biodiesel. Miao and Wu [91] also introduced a new method integrating transesterification process and bio-engineering together for obtaining biodiesel from microalgal oil. Herein, high content of lipid was extracted from the heterotrophic growth of Chlorella protothecoides and n-hexane was used to extract high quantity of microalgal oil from C. protothecoides. It was noticed that the combined method was a successful approach for developing biodiesel from microalgal oil. Wei *et al.* [92] demonstrated the production of biofuels by introducing an approach for developing effective and proper utilization of cellulosic carbons. This strategy was implemented by combining the fermentation route of xylose and pentose as well as acetic acid reduction route into a Saccharomyces cerevisiae strain. With the combination of synthetic biological strategies and different heterologous pathways into a single microbial strain, potential effects were generated for supplementing the production of biofuels. The metabolic routes of simultaneous consumption of cellobiose, xylose and acetic acid are demonstrated in Figure 1.1.

Sarin *et al.* [93] studied the physico-chemical characteristics of Jatropha-Palm biodiesel blends. In this study, the authors discussed the requirement of blending of these two biodiesels as feedstock required for producing biodiesel must exhibit proper integration of saturated and unsaturated fatty compounds for obtaining substantial oxidation stability and low temperature properties. Biodiesel extracted from Jatropha exhibited inadequate oxidation stability and effective low temperature attributes. On the other hand, substantial oxidative stability and poor low temperature properties were exhibited by biodiesel extracted from Palm. Hence, Jatropha and Palm biodiesels were blended for obtaining biodiesel which exhibited substantial oxidation stability and low temperature properties. Additionally, the authors discussed the issues associated with the stability of

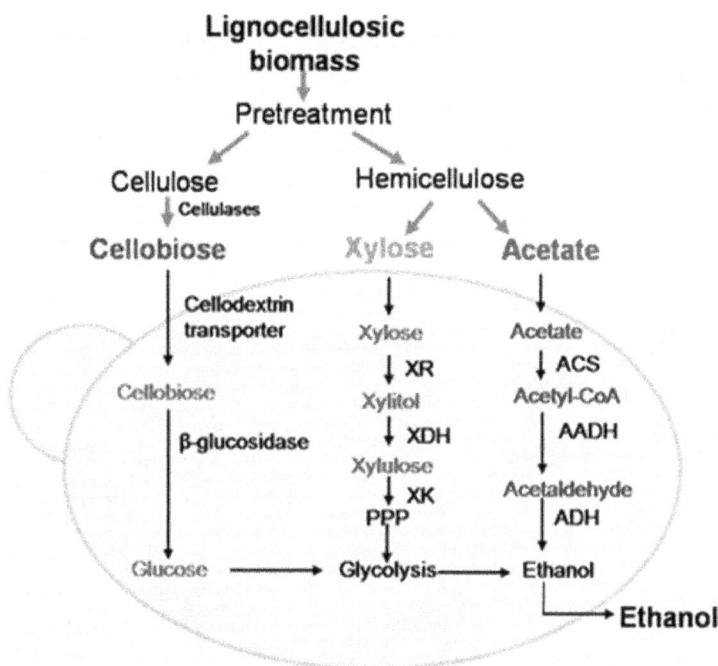

Figure 1.1 Schematic diagram for producing biofuel through simultaneous consumption of cellobiose, xylose, and acetic acid from lignocellulosic biomass by yeast. Reproduced from Reference 92 with permission from American Chemical Society.

biodiesel as it needs antioxidant for the storage purposes. According to EN 14112 specification, it is essential to add 200 mg/L of antioxidant into biodiesel for maintaining its stability, which is comparatively higher than the antioxidant concentration required in petroleum-extracted diesel. For reducing the dosage of antioxidant, Palm biodiesel was blended with Jatropha biodiesel in an appropriate proportion as Palm biodiesel does not require the presence of antioxidant for maintaining its own stability. It was noticed that dosage of antioxidant was decreased by 80-90% after Palm biodiesel was mixed with Jatropha biodiesel at a concentration of 20-40%. Also, the disadvantages associated with Palm biodiesel, such as inappropriate cloud point and pour point, were neutralized by Jatropha biodiesel. Thus, optimized integration of two biodiesels could lead to substantial oxidation stability and low temperature property. In another study, Pal *et al.* [94] synthesized a series of Mn-doped mesoporous ceria-silica composites and utilized these as heterogeneous and reusable catalysts for liquid phase transesterification of different esters (containing both long chain and short chain alcohols) in the presence of different alcohols like n-octanol, n-octanol, etc., for the generation of biodiesel under mild conditions. Transmission electron microscopy (TEM) revealed the hexagonal arrangement of pores in the composites. Scanning electron microscopy (SEM) also revealed that the particles sizes of these composites were in the range 300-400 nm (Figure 1.2).

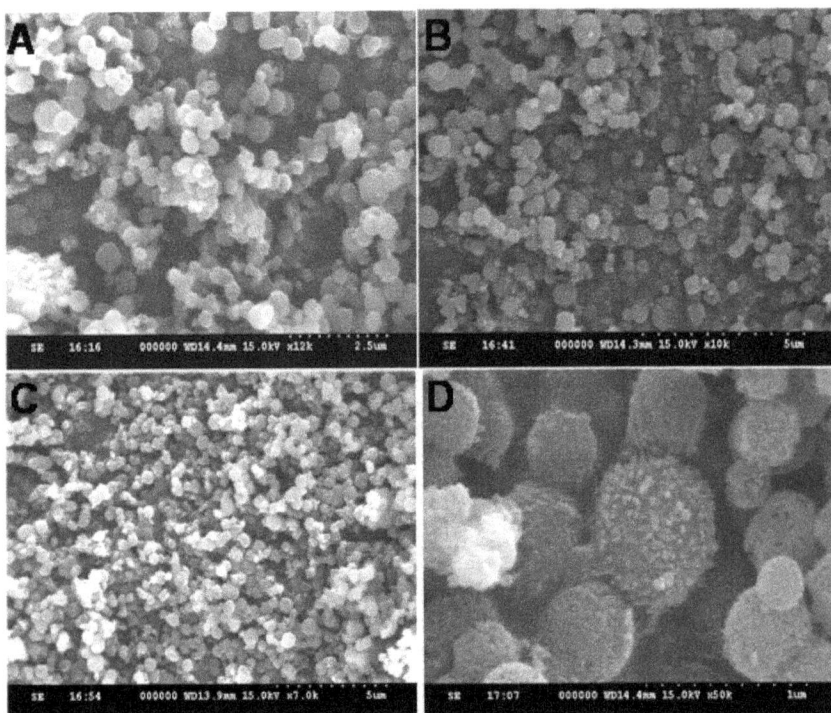

Figure 1.2 Scanning electron microscopy images of the Mn-doped ceria-silica composites. Reproduced from Reference 94 with permission from American Chemical Society.

Sahoo and Das [95] reported different methods of producing biodiesel from Polanga, Jatropha and Karanja oils. The authors employed one-step base catalyzed transesterification process for Karanja and Jatropha oils for the production of biodiesel, while a three-step acid base process was adopted for Polanga oil. In another study, Neumann *et al.* [96] introduced fully off-grid approach for producing cellulosic bioethanol which involves the integration of solar steam processing and solar distillation. Basically, three steps are followed for the development of cellulosic bioethanol: a) sugars are extracted from cellulosic feedstock, b) ethanol is obtained from the fermentation of sugars and c) distillation process is used for ethanol purification. In comparison, the conventional techniques involving extraction and distillation stages for producing bioethanol are associated with consumption of high energy, thereby, enhancing the cost. In this study, nanoparticle-activated solar steam was generated by the illumination of light-absorbing nanoparticles (dispersed in water) with sunlight. With the direct exposure of solar irradiation, ethanol could be obtained from cellulosic feedstock. The authors constructed the system based on nanoparticle-enabled solar steam generation where it was possible to deliver steam into a container of 7.5 L at the temperature ranging from 150-180°C for degrading the cellulosic chains of feedstock into monomeric sugars.

Schematic demonstration as well as digital images of the three-step procedure for generating solar biofuel are demonstrated in Figure 1.3.

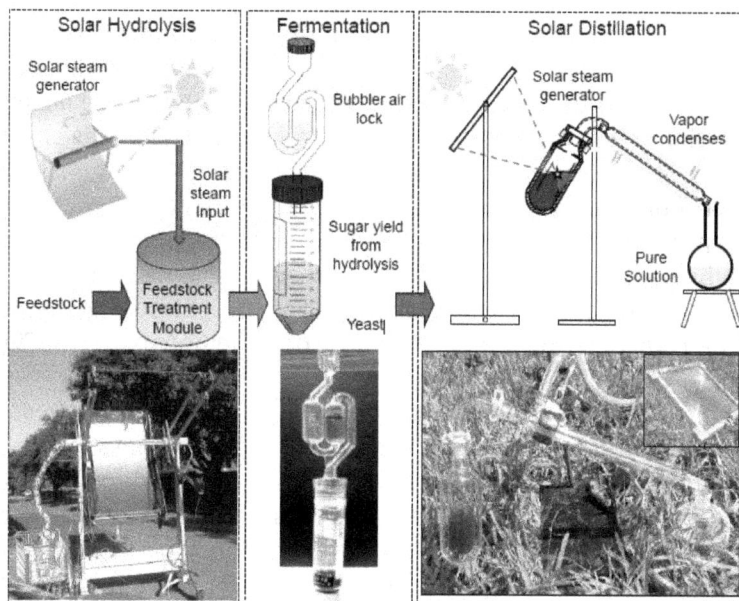

Figure 1.3 Schematic demonstration and digital images of three-step procedure for generating solar biofuel. Reproduced from Reference 96 with permission from American Chemical Society.

Mu *et al.* [97] investigated the integrated bioprocess coupling preparation of biodiesel by lipase with microbial production of 1,3-propanediol using a hollow fiber membrane. During the transesterification process, glycerol was released as a byproduct which could flow through the membrane and transformed into 1,3-propanediol by Klebsiella pneumonia. After the reaction time of 20 h, 84% of biodiesel was generated. It was also observed that productivity of 1,3-propanediol was 1.7 $g.L^{-1}.h^{-1}$ and the value of molar yield of 1,3-propanediol was observed as 0.47 $mol.mol^{-1}$. Glycerol inhibition was avoided by the integrated bioprocess, thereby, decreasing the cost of production and enhancing the yield of biodiesel and 1,3-propanediol. In another study, Yuan *et al.* [98] reported the development of biodiesel from waste rapeseed oil (feedstock). Response surface methodology was employed for optimizing the conditions in order to achieve the highest transformation to biodiesel. Gas chromatography/mass spectroscopy was used for analyzing the generated biodiesel and the as-prepared biodiesel was observed to contain six fatty acid methyl esters. Dorado *et al.* [99] also conducted the studies for optimizing the alkali catalyzed transesterification process of Brassica carinata oil. In order to obtain an economical procedure for generating biodiesel, the reaction was carried out with 30 min stirring of a mixture of oil/methanol (in a ratio of 1:4.6) and potassium hydroxide in the temperature range 20-45 °C. These conditions were observed to be suitable for the low-cost production of biodiesel. Gas chromatography analysis was carried out for determining the composition of fatty acid of B. carinata oil and its esters. The results disclosed that the content of free fatty acid increased due to the presence of erucic acid which inhibited the transformation of B. carinata oil into its methyl ester. Schematic representation of the biofuel generation through transesterification process has been demonstrated in Figure 1.4.

Johnson and Wen [100] reported the generation of biodiesel by applying the direct transesterification of algal biomass. Microalga Schizochytrium limacinum was utilized for the production of biodiesel as it generates high levels of biomass and total fatty acid. Biodiesel generated from direct transesterification process fulfilled the ASTM specifications, thereby, representing a viable biodiesel product. In another study, Alexander *et al.* [101] reported the generation of biofuels using fish oil. For this purpose, the authors carried out the partial hydrogenation of fish oil extracted fatty acid methyl esters (FAME) using a batch reactor at room temperature for < 15 minutes.

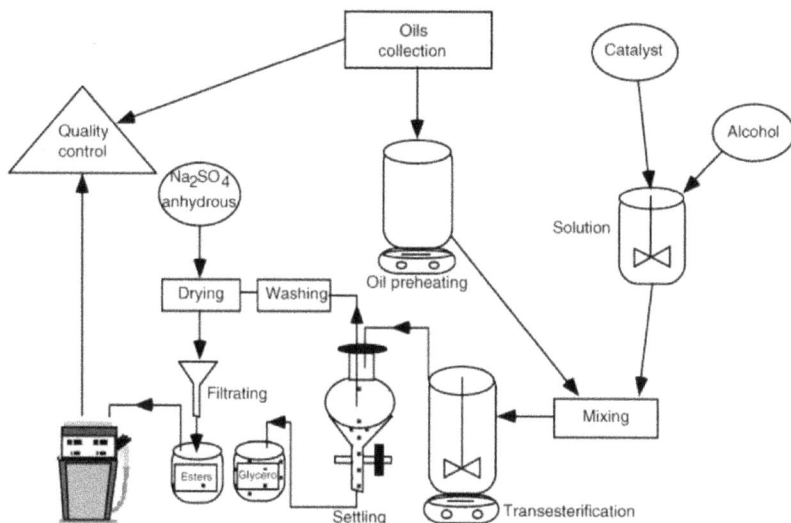

Figure 1.4 Schematic representation of generating biofuel through transesterification process. Reproduced from Reference 99 with permission from American Chemical Society.

Herein, selective hydrogenation of polyunsaturated fatty acids was carried out, whereas the content of saturated fatty acids remained at a constant level as compared to the feedstock. The as-prepared biofuels and their blends with petroleum-extracted fuel were examined for fuel characteristics as well as for EN 14214 and EN 590 specifications. The results suggested that the obtained biofuels were suitable for generating blends with petroleum-derived fuel and the blends fulfilled EN 590 specification. In a recent study, Phelan *et al.* [102] introduced terpene biosynthesis in Streptomyces cerevisiae for the generation of modified biofuel precursor bisabolene (Figure 1.5). Morshed *et al.* [103] also reported rubber seed oil as an effective feedstock for producing biodiesel. Techniques like cold percolation and mechanical press with and without solvent were employed for the extraction of oil from the rubber seeds. Biodiesel was produced by using the method involving three steps, i.e., oil saponification, soap acidification and esterification of free fatty acid. FAME was generated in the esterification step. The maximum productivity of free fatty acid from rubber seed oil was observed as 86%. Nuclear magnetic resonance (NMR) spectroscopy was also used to analyze the rubber seed oil and biodiesel samples, which confirmed the transformation of rubber seed oil into biodiesel.

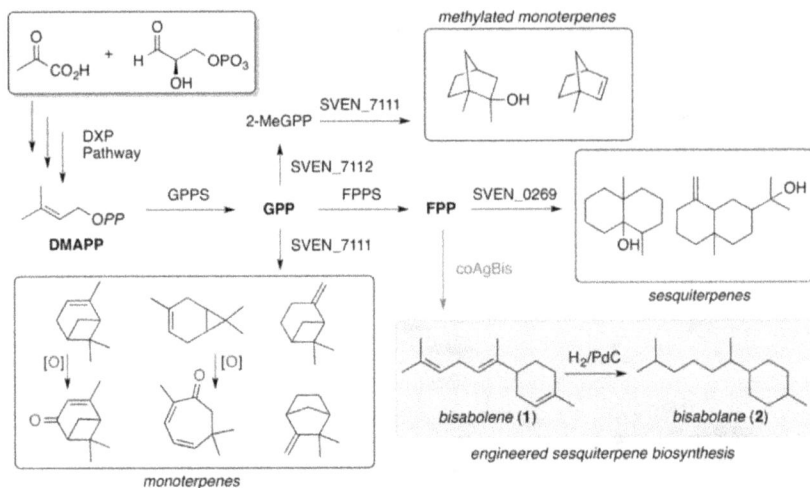

Figure 1.5 Biosynthetic routes to predominant isoprenoids produced by endogenous (black arrows) and heterologous (red arrow) terpene synthases in S. venezuelae. Reproduced from Reference 102 with permission from American Chemical Society.

In another study, Yang *et al.* [104] also reported the production of biodiesel utilizing rubber seed oil as feedstock. The authors developed poly(sodium acrylate) and sodium hydroxide (NaOH-NaPAA) integrated system acting as a water resistant catalyst, which further reduced the unfavorable effects of water content on the base catalyzed transesterification process for producing biodiesel. The results revealed that the modified catalyst system exhibited significant water resistance during transesterification process and maintained high catalytic activity. Ghadge and Raheman [105] reported optimized design for producing biodiesel from mahua oil. In the study, central composite rotatable pattern was utilized for examining the effects of quantity of methanol and acid as well as durability of reaction for reducing free fatty acids of mahua oil during its pre-treatment for the generation of biodiesel. Ge *et al.* [106] also reported an approach to recover microalgae from water for the generation of biofuel. In this study, modification of CO_2-switchable crystalline nanocellulose (CNC) was achieved with 1-(3-aminopropyl)-imidazole (APIm) so that it behaves as a reversible coagulant for the purpose of microalgae harvesting. Positively charged APIm-modified CNC exhibited good dispersion in the carbonated water as compared to unmodified CNC. On CO_2-treatment, the modified CNC was observed to exhibit substantial electrostatic interactions with negatively charged Chlorella vulgaris. Multiple mechanisms were integrated in order to harvest microalgae such as electrostatic interaction during CO_2 sparging stage, enmeshment during air sparging triggered interaction of APIm modified CNC with Chlorella vulgaris, followed by further coagulation, flocculation and sedimentation stages (Figure 1.6). This approach was significant as it involved biocompatible, readily available and natural components like air, CO_2 and cellulose for generating biofuel.

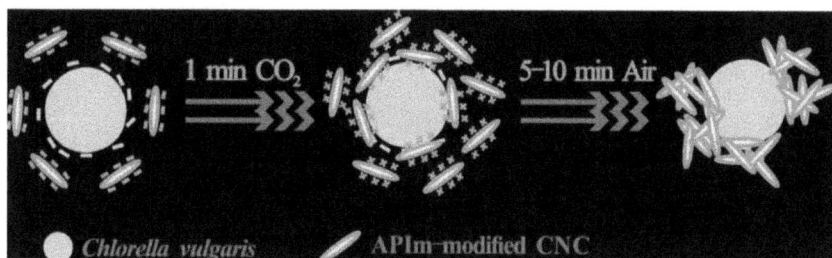

Figure 1.6 Schematic representation of the electrostatic attraction and enmeshment mechanisms involved in interactions between C. vulgaris and APIm-modified CNC. Reproduced from Reference 106 with permission from American Chemical Society.

Xie *et al.* [107] developed Guerbet-type process catalyzed using complexes of ruthenium pincer for generating biofuel from ethanol (Figure 1.7). In another study, Hums *et al.* [108] also performed the life cycle assessment of biodiesel obtained from grease trap waste. Basically, life cycle assessment was utilized to measure the quantity

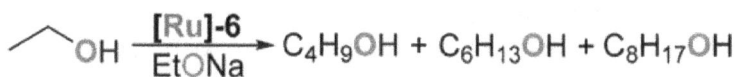

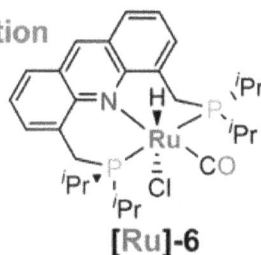

Figure 1.7 Representation of the Guerbet-type process. Reproduced from Reference 107 with permission from American Chemical Society.

of greenhouse gas emissions, necessity of fossil energy and air pollutant emissions for the process of grease trap waste extracted biodiesel. Herein, Monte Carlo simulations were performed for analyzing the sensitivity to the concentration of lipid present in grease trap waste. The assessment displayed complete life cycle of biofuel from the point of arrangement of feedstock (grease trap waste) to the combustion of biofuel (Figure 1.8).

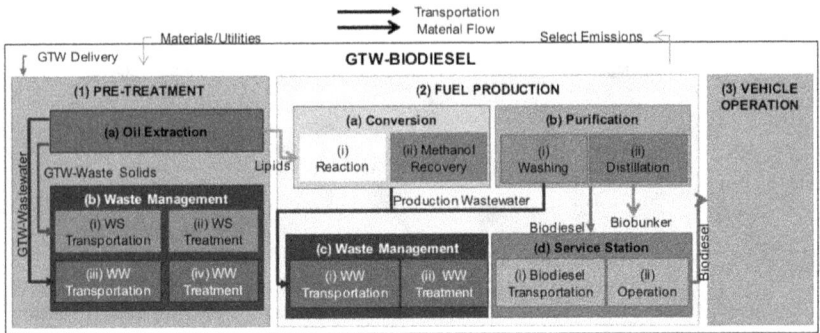

Figure 1.8 System boundary for the grease trap waste (GTW)-biodiesel process. Each of the three main stages include the material and energy inputs and emission outputs for (1) pretreatment, (2) fuel production and (3) vehicle operation. Reproduced from Reference 108 with permission from American Chemical Society.

1.6 Conclusions

Biofuels are beneficial alternatives to the petroleum-derived fuels. These fuels exhibit enhanced sustainability and reduction in the level of greenhouse gas emissions. In addition, the existing engines or vehicles do not require any special modification for utilizing biofuels. Overcoming the challenges associated with productivity and cost in the near future will ensure high demand of bio-derived fuels owing to their beneficial features.

References

1. Bringe, N. A., (2005) Soybean oil composition for biodiesel. In: *The Biodiesel Handbook*, Knothe, G. and Van Gerpen (eds.), AOCS Publishing, USA.

2. Pahl, G. (2008) *Biodiesel: Growing a New Energy Economy*, 2nd edition, Chelsea Green Publishing, USA.
3. Van Gerpen, J. (2005) Biodiesel processing and production. *Fuel Processing Technology*, **86**, 1097-1107.
4. Ma, F., Clements, L. D., and Hanna, M. A. (1998) Biodiesel fuel from animal fat. Ancillary studies on transesterification of beef tallow. *Industrial & Engineering Chemistry Research*, **37**, 3768-3771.
5. Haas, M. J., McAloon, A. J., Yee, W. C., and Foglia, T. A. (2006) A process model to estimate biodiesel production costs. *Bioresource Technology*, **97**, 671-678.
6. Bender, M. (1999) Economic feasibility review for community-scale farmer cooperatives for biodiesel. *Bioresource Technology*, **70**, 81-87.
7. Korbitz, W. (1999) Biodiesel production in Europe and North America, an encouraging prospect. *Renewable Energy*, **16**, 1078-1083.
8. Bozbas, K. (2008) Biodiesel as an alternative motor fuel: production and policies in the European Union. *Renewable and Sustainable Energy Reviews*, **12**, 542-552.
9. Dorado, M., Cruz, F., Palomar, J., and Lopez, F. (2006) An approach to the economics of two vegetable oil-based biofuels in Spain. *Renewable Energy*, **31**, 1231-1237.
10. Tyson, K. S., and McCormick, R. (2001) Biodiesel Handling and Use Guide. *National Renewable Energy Laboratory*, USA. Online: http://biodiesel.org/docs/using-hotline/nrel-handling-and-use.pdf?sfvrsn=4 (accessed 12th June 2018).
11. Hoekman, S. K. (2009) Biofuels in the US - challenges and opportunities. *Renewable Energy*, **34**, 14-22.
12. Larson, E. D. (2008) Biofuel Production Technologies: Status, Prospects and Implications for Trade and Development. *United Nations Conference on Trade and Development*, United Nations. Online: http://unctad.org/en/Docs/ditcted200710 en.pdf (accessed 11th June 2018).
13. Patil, V., Tran, K.-Q., and Giselrod, H. R. (2008) Towards sustainable production of biofuels from microalgae. *International Journal of Molecular Sciences*, **9**, 1188-1195.
14. Gibbons, W. R., and Westby, C. A. (1989) Cofermentation of sweet sorghum juice and grain for production of fuel ethanol and distillers' wet grain. *Biomass*, **18**, 43-57.
15. Suresh, K., and Rao, L. V. (1999) Utilization of damaged sorghum and rice grains for ethanol production by simultaneous saccharification and fermentation. *Bioresource Technology*, **68**, 301-304.
16. Turhollow, A. F., and Heady, E. O. (1986) Large-scale ethanol production from corn and grain sorghum and improving conversion technology. *Energy in Agriculture*, **5**, 309-316.
17. Zhao, R., Bean, S., Wang, D., Park, S., Schober, T., and Wilson, J.

(2009) Small-scale mashing procedure for predicting ethanol yield of sorghum grain. *Journal of Cereal Science*, **49**, 230-238.

18. Love, G., Gough, S., Brady, D., Barron, N., Nigam, P., Singh, D., Marchant, R., and McHale, A. (1998) Continuous ethanol fermentation at 45 °C using Kluyveromyces marxianus IMB3 immobilized in Calcium alginate and kissiris. *Bioprocess Engineering*, **18**, 187-189.

19. Nigam, P., Banat, I., Singh, D., McHale, A., and Marchant, R. (1997) Continuous ethanol production by thermotolerant Kluyveromyces marxianus IMB3 immobilized on mineral Kissiris at 45 C. *World Journal of Microbiology and Biotechnology*, **13**, 283-288.

20. Brady, D., Nigam, P., Marchant, R., and McHale, A. (1997) Ethanol production at 45 C by alginate-immobilized Kluyveromyces marxianus IMB3 during growth on lactose-containing media. *Bioprocess Engineering*, **16**, 101-104.

21. Brady, D., Nigam, P., Marchant, R., Singh, D., and McHale, A. (1997) The effect of Mn^{2+} on ethanol production from lactose using Kluyveromyces marxianus IMB3 immobilized in magnetically responsive matrices. *Bioprocess and Biosystems Engineering*, **17**, 31-34.

22. Riordan, C., Love, G., Barron, N., Nigam, P., Marchant, R., McHale, L., and McHale, A. (1996) Production of ethanol from sucrose at 45 °C by alginate-immoblized preparations of the thermotolerant yeast strain Kluyveromyces marxianus IMB3. *Bioresource Technology*, **55**, 171-173.

23. Love, G., Nigam, P., Barron, N., Singh, D., Marchant, R., and McHale, A. (1996) Ethanol production at 45 °C using preparations of Kluyveromyces marxianus IMB3 immobilized in calcium alginate and kissiris. *Bioprocess and Biosystems Engineering*, **15**, 275-277.

24. Banat, I. M., Nigam, P., and Marchant, R. (1992) Isolation of thermotolerant, fermentative yeasts growing at 52 C and producing ethanol at 45 °C and 50 °C. *World Journal of Microbiology and Biotechnology*, **8**, 259-263.

25. Thegarid, N., Fogassy, G., Schuurman, Y., Mirodatos, C., Stefanidis, S., Iliopoulou, E., Kalogiannis, K., and Lappas, A. (2014) Second-generation biofuels by co-processing catalytic pyrolysis oil in FCC units. *Applied Catalysis B: Environmental*, **145**, 161-166.

26. Schenk, P. M., Thomas-Hall, S. R., Stephens, E., Marx, U. C., Mussgnug, J. H., Posten, C., Kruse, O., and Hankamer, B. (2008) Second generation biofuels: high-efficiency microalgae for biodiesel production. *Bioenergy Research*, **1**, 20-43.

27. Dias, M. O., Junqueira, T. L., Cavalett, O., Cunha, M. P., Jesus, C. D., Rossell, C. E., Maciel Filho, R., and Bonomi, A. (2012) Integrated versus stand-alone second generation ethanol production from sugarcane bagasse and trash. *Bioresource Technology*, **103**, 152-161.

28. Brennan, L., and Owende, P. (2010) Biofuels from microalgae - a review of technologies for production, processing, and extractions of

biofuels and co-products. *Renewable and Sustainable Energy Reviews*, **14**, 557-577.

29. Carere, C. R., Sparling, R., Cicek, N., and Levin, D. B. (2008) Third generation biofuels via direct cellulose fermentation. *International Journal of Molecular Sciences*, **9**, 1342-1360.

30. Saidane-Bchir, F., El Falleh, A., Ghabbarou, E., and Hamdi, M. (2016) 3rd generation bioethanol production from microalgae isolated from slaughterhouse wastewater. *Waste and Biomass Valorization*, **7**, 1041-1046.

31. Zhu, L., Zong, M., and Wu, H. (2008) Efficient lipid production with Trichosporonfermentans and its use for biodiesel preparation. *Bioresource Technology*, **99**, 7881-7885.

32. Samori, C., Torri, C., Samorì, G., Fabbri, D., Galletti, P., Guerrini, F., Pistocchi, R., and Tagliavini, E. (2010) Extraction of hydrocarbons from microalga Botryococcus braunii with switchable solvents. *Bioresource Technology* **101**, 3274-3279.

33. El-Shimi, H., Attia, N. K., El-Sheltawy, S., and El-Diwani, G. (2013) Biodiesel production from Spirulina-platensis microalgae by in-situ transesterification process. *Journal of Sustainable Bioenergy Systems*, **3**, 224-233.

34. Chng, L. M., Chan, D. J., and Lee, K. T. (2016) Sustainable production of bioethanol using lipid-extracted biomass from Scenedesmus dimorphus. *Journal of Cleaner Production*, **130**, 68-73.

35. Abreu, A. P., Fernandes, B., Vicente, A. A., Teixeira, J., and Dragone, G. (2012) Mixotrophic cultivation of Chlorella vulgaris using industrial dairy waste as organic carbon source. *Bioresource Technology*, **118**, 61-66.

36. Srinophakun, P., Thanapimmetha, A., Rattanaphanyapan, K., Sahaya, T., and Saisriyoot, M. (2017) Feedstock production for third generation biofuels through cultivation of Arthrobacter AK19 under stress conditions. *Journal of Cleaner Production*, **142**, 1259-1266.

37. Wang, Y., Wu, H., and Zong, M. (2008) Improvement of biodiesel production by lipozyme TL IM-catalyzed methanolysis using response surface methodology and acyl migration enhancer. *Bioresource Technology*, **99**, 7232-7237.

38. Vicente, G., Martınez, M., and Aracil, J. (2004) Integrated biodiesel production: a comparison of different homogeneous catalysts systems. *Bioresource Technology*, **92**, 297-305.

39. Antolın, G., Tinaut, F., Briceno, Y., Castano, V., Perez, C., and Ramırez, A. (2002) Optimisation of biodiesel production by sunflower oil transesterification. *Bioresource Technology*, **83**, 111-114.

40. Lang, X., Dalai, A. K., Bakhshi, N. N., Reaney, M. J., and Hertz, P. (2001) Preparation and characterization of bio-diesels from various bio-oils. *Bioresource Technology*, **80**, 53-62.

41. Pramanik, K. (2003) Properties and use of Jatropha curcas oil and

diesel fuel blends in compression ignition engine. *Renewable Energy*, **28**, 239-248.

42. Meher, L. C., Kulkarni, M. G., Dalai, A. K., and Naik, S. N. (2006) Transesterification of karanja (Pongamia pinnata) oil by solid basic catalysts. *European Journal of Lipid Science and Technology*, **108**, 389-397.

43. Abbaszaadeh, A., Ghobadian, B., Omidkhah, M. R., and Najafi, G. (2012) Current biodiesel production technologies: a comparative review. *Energy Conversion and Management*, **63**, 138-148.

44. Bengisu, G. (2014) Alternatif yakıt kaynağı olarak biyoetanol/Bioethanol as an alternative fuel source. *Alınteri Zirai Bilimler Dergisi*, **27**, 43-52.

45. The State of Food and Agriculture, Biofuels: Prospects, Risks and Opportunities (2008). *Food and Agriculture Organization of the United Nations*, Italy. Online: http://www.fao.org/3/a-i0100e.pdf (assessed 9th June 2018).

46. Sustainable Biofuels: Prospects and Challenges (2008). *The Royal Society*, UK. Online: https://royalsociety.org/~/media/royal society content/policy/publications/2008/7980.pdf (assessed 1st June 2018).

47. Verma, G., Nigam, P., Singh, D., and Chaudhary, K. (2000) Bioconversion of starch to ethanol in a single-step process by coculture of amylolytic yeasts and Saccharomyces cerevisiae 21. *Bioresource Technology*, **72**, 261-266.

48. Singh, D., Dahiya, J. S., and Nigam, P. (1995) Simultaneous raw starch hydrolysis and ethanol fermentation by glucoamylase from Rhizoctonia solani and Saccharomyces cerevisiae. *Journal of Basic Microbiology*, **35**, 117-121.

49. Sindhu, R., Kuttiraja, M., Binod, P., Sukumaran, R.K., and Pandey, A. (2014) Bioethanol production from dilute acid pretreated Indian bamboo variety (Dendrocalamus sp.) by separate hydrolysis and fermentation. *Industrial Crops and Products*, **52**, 169-176.

50. Trivedi, N., Gupta, V., Reddy, C., and Jha, B. (2013) Enzymatic hydrolysis and production of bioethanol from common macrophytic green alga Ulva fasciata Delile. *Bioresource Technology*, **150**, 106-112.

51. Kwon, E. E., Yi, H., and Jeon, Y. J. (2013) Sequential co-production of biodiesel and bioethanol with spent coffee grounds. *Bioresource Technology*, **136**, 475-480.

52. Enquist-Newman, M., Faust, A. M. E., Bravo, D. D., Santos, C. N. S., Raisner, R. M., Hanel, A., Sarvabhowman, P., Le, C., Regitsky, D. D., and Cooper, S. R. (2014) Efficient ethanol production from brown macroalgae sugars by a synthetic yeast platform. *Nature*, **505**, 239-243.

53. Kumar, S., Gupta, R., Kumar, G., Sahoo, D., and Kuhad, R. C. (2013) Bioethanol production from Gracilaria verrucosa, a red alga, in a bi-

orefinery approach. *Bioresource Technology*, **135**, 150-156.

54. Borines, M. G., de Leon, R. L., and Cuello, J. L. (2013) Bioethanol production from the macroalgae Sargassum spp. *Bioresource Technology*, **138**, 22-29.

55. Shukla, R., Kumar, M., Chakraborty, S., Gupta, R., Kumar, S., Sahoo, D., and Kuhad, R.C. (2016) Process development for the production of bioethanol from waste algal biomass of Gracilaria verrucosa. *Bioresource Technology*, **220**, 584-589.

56. Raghavi, S., Sindhu, R., Binod, P., Gnansounou, E., and Pandey, A. (2016) Development of a novel sequential pretreatment strategy for the production of bioethanol from sugarcane trash. *Bioresource Technology*, **199**, 202-210.

57. Klein, M., Griess, O., Pulidindi, I. N., Perkas, N., and Gedanken, A. (2016) Bioethanol production from Ficus religiosa leaves using microwave irradiation. *Journal of Environmental Management*, **177**, 20-25.

58. Ramey, D. (2004) Butanol. *The Light Party*, USA. Online: http://www.lightparty.com/Energy/Butanol.html (assessed 2nd June 2018).

59. Biobutanol, European Biofuels Technology Platform (2009). Online: http://biofuelstp.eu/newsletters/newsletter_archive.html# (assessed 1st July 2018).

60. Wu, M., Wang, M., Liu, J., and Huo, H. (2007) Life-cycle Assessment of Corn-based Butanol as a Potential Transportation Fuel. *Argonne National Laboratory*, USA. Online: https://anl.box.com/s/s11ha0ko xgy4rme4ykgkxw5yrz1omofs (assessed 5th July 2018).

61. Qureshi, N., Saha, B. C., Dien, B., Hector, R. E., and Cotta, M. A. (2010) Production of butanol (a biofuel) from agricultural residues: Part I - Use of barley straw hydrolysate. *Biomass and Bioenergy*, **34**, 559-565.

62. Jiang, W., Zhao, J., Wang, Z., and Yang, S.-T. (2014) Stable high-titer n-butanol production from sucrose and sugarcane juice by Clostridium acetobutylicum JB200 in repeated batch fermentations. *Bioresource Technology*, **163**, 172-179.

63. Moradi, F., Amiri, H., Soleimanian-Zad, S., Ehsani, M. R., and Karimi, K. (2013) Improvement of acetone, butanol and ethanol production from rice straw by acid and alkaline pretreatments. *Fuel*, **112**, 8-13.

64. Du, T.-f., He, A.-y., Wu, H., Chen, J.-n., Kong, X.-p., Liu, J.-l., Jiang, M., and Ouyang, P.-k. (2013) Butanol production from acid hydrolyzed corn fiber with Clostridium beijerinckii mutant. *Bioresource Technology*, **135**, 254-261.

65. Xin, F., Basu, A., Yang, K.-L., and He, J. (2016) Strategies for production of butanol and butyl-butyrate through lipase-catalyzed esterification. *Bioresource Technology*, **202**, 214-219.

66. Kong, X., Xu, H., Wu, H., Wang, C., He, A., Ma, J., Ren, X., Jia, H., Wei, C.,

and Jiang, M. (2016) Biobutanol production from sugarcane bagasse hydrolysate generated with the assistance of gamma-valerolactone. *Process Biochemistry*, **51**, 1538-1543.

67. Houweling-Tan, B., Sperber, B. L., van der Wal, H., Bakker, R. R., and Lopez-Contreras, A. M. (2016) Barley dried distillers grains and solubles (DDGS) as feedstock for production of acetone, butanol and ethanol. *BAOJ Microbio*, **2**, 013.

68. Sharma, Y., Singh, B., and Upadhyay, S. (2008) Advancements in development and characterization of biodiesel: A review. *Fuel*, **87**, 2355-2373.

69. Schuchardt, U., Sercheli, R., and Vargas, R. M. (1998) Transesterification of vegetable oils: a review. *Journal of the Brazilian Chemical Society*, **9**, 199-210.

70. Wang, L., and Yang, J. (2007) Transesterification of soybean oil with nano-MgO or not in supercritical and subcritical methanol. *Fuel*, **86**, 328-333.

71. Kouzu, M., and Hidaka, J.-s. (2012) Transesterification of vegetable oil into biodiesel catalyzed by CaO: A review. *Fuel*, **93**, 1-12.

72. Dalai, A. K., Kulkarni, M. G., and Meher, L. C. (2006) Biodiesel Productions from Vegetable Oils Using Heterogeneous Catalysts and Their Applications as Lubricity Additives. 2006 IEEE EIC Climate Change Conference, Canada, pp. 1-8 (doi: 10.1109/EICCCC.2006.27 7228.

73. Jitputti, J., Kitiyanan, B., Rangsunvigit, P., Bunyakiat, K., Attanatho, L., and Jenvanitpanjakul, P. (2006) Transesterification of crude palm kernel oil and crude coconut oil by different solid catalysts. *Chemical Engineering Journal*, **116**, 61-66.

74. Patil, P. D., and Deng, S. (2009) Optimization of biodiesel production from edible and non-edible vegetable oils. *Fuel*, **88**, 1302-1306.

75. Singh, A., He, B., Thompson, J., and Van Gerpen, J. (2006) Process optimization of biodiesel production using alkaline catalysts. *Applied Engineering in Agriculture*, **22**, 597-600.

76. Karmakar, A., Karmakar, S., and Mukherjee, S. (2010) Properties of various plants and animals feedstocks for biodiesel production. *Bioresource technology*, **101**, 7201-7210.

77. Canakci, M., and Van Gerpen, J. (1999) Biodiesel production via acid catalysis. *Transactions of the ASAE-American Society of Agricultural Engineers*, **42**, 1203-1210.

78. Vasudevan, P. T., and Briggs, M. (2008) Biodiesel production - current state of the art and challenges. *Journal of Industrial Microbiology & Biotechnology*, **35**, 421.

79. Tan, T., Lu, J., Nie, K., Deng, L., and Wang, F. (2010) Biodiesel production with immobilized lipase: a review. *Biotechnology Advances*, **28**, 628-634.

80. Tan, K. T., Lee, K. T., and Mohamed, A. R. (2009) Production of FAME

by palm oil transesterification via supercritical methanol technology. *Biomass and Bioenergy*, **33**, 1096-1099.

81. Encinar, J.M., González, J.F., Sabio, E., and Ramiro, M.J. (1999) Preparation and properties of biodiesel from Cynara C ardunculus L. oil. *Industrial & Engineering Chemistry Research*, **38**, 2927-2931.

82. van Gerpen, J., Shanks, B., Pruszko, R., Clements, D., and Knothe, G. (2004) Biodiesel analytical methods. *National Renewable Energy Laboratory*, USA. Online: http://www.graham-laming.com/bd/testing_biodiesel.pdf (assessed 10th July 2018).

83. Zheng, S., Kates, M., Dube, M., and McLean, D. (2006) Acid-catalyzed production of biodiesel from waste frying oil. *Biomass and Bioenergy*, **30**, 267-272.

84. Demirbas, A., and Demirbas, M. F. (2011) Importance of algae oil as a source of biodiesel. *Energy Conversion and Management*, **52**, 163-170.

85. Leung, D., and Guo, Y. (2006) Transesterification of neat and used frying oil: optimization for biodiesel production. *Fuel Processing Technology*, **87**, 883-890.

86. Rashid, U., and Anwar, F. (2008) Production of biodiesel through optimized alkaline-catalyzed transesterification of rapeseed oil. *Fuel*, **87**, 265-273.

87. Dube, M., Tremblay, A., and Liu, J. (2007) Biodiesel production using a membrane reactor. *Bioresource Technology*, **98**, 639-647.

88. Berchmans, H. J., and Hirata, S. (2008) Biodiesel production from crude Jatropha curcas L. seed oil with a high content of free fatty acids. *Bioresource Technology*, **99**, 1716-1721.

89. Kouzu, M., Kasuno, T., Tajika, M., Sugimoto, Y., Yamanaka, S., and Hidaka, J. (2008) Calcium oxide as a solid base catalyst for transesterification of soybean oil and its application to biodiesel production. *Fuel*, **87**, 2798-2806.

90. Liu, X., Piao, X., Wang, Y., Zhu, S., and He, H. (2008) Calcium methoxide as a solid base catalyst for the transesterification of soybean oil to biodiesel with methanol. *Fuel*, **87**, 1076-1082.

91. Miao, X., and Wu, Q. (2006) Biodiesel production from heterotrophic microalgal oil. *Bioresource Technology*, **97**, 841-846.

92. Wei, N., Oh, E. J., Million, G., Cate, J. H., and Jin, Y.-S. (2015) Simultaneous utilization of cellobiose, xylose, and acetic acid from lignocellulosic biomass for biofuel production by an engineered yeast platform. *ACS Synthetic Biology*, **4**, 707-713.

93. Sarin, R., Sharma, M., Sinharay, S., and Malhotra, R. K. (2007) Jatropha–palm biodiesel blends: an optimum mix for Asia. *Fuel*, **86**, 1365-1371.

94. Pal, N., Cho, E.-B., Kim, D., and Jaroniec, M. (2014) Mn-doped ordered mesoporous ceria-silica composites and their catalytic properties toward biofuel production. *The Journal of Physical Chemistry*

C, **118**, 15892-15901.

95. Sahoo, P., and Das, L. (2009) Process optimization for biodiesel production from Jatropha, Karanja and Polanga oils. *Fuel*, **88**, 1588-1594.

96. Neumann, O., Neumann, A. D., Tian, S., Thibodeaux, C., Shubhankar, S., Müller, J., Silva, E., Alabastri, A., Bishnoi, S. W., and Nordlander, P. (2016) Combining solar steam processing and solar distillation for fully off-grid production of cellulosic bioethanol. *ACS Energy Letters*, **2**, 8-13.

97. Mu, Y., Xiu, Z.-L., and Zhang, D.-J. (2008) A combined bioprocess of biodiesel production by lipase with microbial production of 1, 3-propanediol by Klebsiella pneumoniae. *Biochemical Engineering Journal*, **40**, 537-541.

98. Yuan, X., Liu, J., Zeng, G., Shi, J., Tong, J., and Huang, G. (2008) Optimization of conversion of waste rapeseed oil with high FFA to biodiesel using response surface methodology. *Renewable Energy*, **33**, 1678-1684.

99. Dorado, M. P., Ballesteros, E., López, F. J., and Mittelbach, M. (2004) Optimization of alkali-catalyzed transesterification of Brassica C arinata oil for biodiesel production. *Energy & Fuels*, **18**, 77-83.

100. Johnson, M. B., and Wen, Z. (2009) Production of biodiesel fuel from the microalga Schizochytrium limacinum by direct transesterification of algal biomass. *Energy & Fuels*, **23**, 5179-5183.

101. Studentschnig, A. F., Schober, S., and Mittelbach, M. (2015) Partial hydrogenation of fish oil methyl esters for the production of biofuels. *Energy & Fuels*, **29**, 3776-3779.

102. Phelan, R. M., Sekurova, O. N., Keasling, J. D., and Zotchev, S. B. (2014) Engineering terpene biosynthesis in Streptomyces for production of the advanced biofuel precursor bisabolene. *ACS Synthetic Biology*, **4**, 393-399.

103. Morshed, M., Ferdous, K., Khan, M. R., Mazumder, M., Islam, M., and Uddin, M. T. (2011) Rubber seed oil as a potential source for biodiesel production in Bangladesh. *Fuel*, **90**, 2981-2986.

104. Yang, R., Su, M., Zhang, J., Jin, F., Zha, C., Li, M., and Hao, X. (2011) Biodiesel production from rubber seed oil using poly (sodium acrylate) supporting NaOH as a water-resistant catalyst. *Bioresource Technology*, **102**, 2665-2671.

105. Ghadge, S. V., and Raheman, H. (2006) Process optimization for biodiesel production from mahua (Madhuca indica) oil using response surface methodology. *Bioresource Technology*, **97**, 379-384.

106. Ge, S., Champagne, P., Wang, H., Jessop, P. G., and Cunningham, M. F. (2016) Microalgae recovery from water for biofuel production using CO_2-switchable crystalline nanocellulose. *Environmental Science & Technology*, **50**, 7896-7903.

107. Xie, Y., Ben-David, Y., Shimon, L. J., and Milstein, D. (2016) Highly eff-

icient process for production of biofuel from ethanol catalyzed by ruthenium pincer complexes. *Journal of the American Chemical Society*, **138**, 9077-9080.

108. Hums, M. E., Cairncross, R. A., and Spatari, S. (2016) Life-cycle assessment of biodiesel produced from grease trap waste. *Environmental Science & Technology*, **50**, 2718-2726.

Chapter 2

Selective Hydrogenation of Furfural and Levulinic Acid to Biofuels

Zixiao Yi,[a] Ruiqi Li,[a] Kai Yan,[a,*] Rongliang Qiu,[a] Shu Wang[b] and Huixia Luo[b,*]

[a]School of Environmental Science and Engineering, Sun Yat-sen University, 135 Xingang Xi Road, Guangzhou 510275, China
[b]School of Material Sciences and Engineering, Sun Yat-sen university, 135 Xingang Xi Road, Guangzhou 510275, China
*Corresponding authors: yank9@mail.sysu.edu.cn; luohx7@mail.sysu.edu.cn

2.1 Introduction

A voluminous body of work over the past several decades has been devoted to searching renewable energy to replace the limited fossil fuel sources and reduce the associated environmental issues [1-6]. Replacing fossil fuels with environmentally-friendly and sustainable alternative sources of energy have exhibited promising aspects over the last decades [7-11]. Among the renewable energy sources, biomass is very attractive due to its wide distribution, low-cost and carbon-neutrality [2,12-16]. It has the burgeoning potential for integration with the current world energy infrastructure. The conversion of biomass into biofuels as well as value-added chemicals can be briefly divided into two periods [2,13,17-19]. The so-called "1st generation" biofuels are biodiesel and bioethanol, where biodiesel has similar properties to conventional diesel fuel in terms of molecular weight and combustion behavior. The production of biodiesel is often achieved through the transesterification of triglycerides from plant oils. Bioethanol, often used as a gasoline additive, is produced through the fermentation of corn, sugarcane and soybeans, etc. For the fermentation process, the original feedstocks including wood, grass and agricultural wastes are difficult to be fermented into ethanol and generated a lot of waste in the process. The feedstocks for "2nd generation" biofuels have a much lower content of simple sugars and are generated from the larger polymeric cellulose, hemicellulose and lignin [20,21]. The high potential of conversion of biomass and biomass based platform chemicals into high value chemicals and biofuels has attracted significant research attention, thereby, leading to the better competitive superiority of these high-valued chemicals and biofuels against fossil-based products [20,22-26]. Current technologies (e.g., gasification, pyrolysis and aqueous phase processing) have been observed to successfully convert biomass into high-value chemicals and biofuels [2,16,27-31].

In this context, top bio-renewable chemicals (Figure 2.1) produced via chemical and biological processes have been identified in recent years. The energy report from Department of Energy (DOE), named "Top Value Added Chemicals from Biomass" [32], has suggested "Top 10" valuable candidates and its update by Bozell *et al.* [33] proposed "Top 10+ 4" valuable chemicals. Along with these candidates, furfural and levulinic acid (LA) have also been considered as two platforms in lignocellulosic biorefineries [16,27,34-37]. These are the dehydration products of carbohydrates (e.g., fructose and xylose), which are derived from lignocellulosic biomass, depicted in Figure 2.2. Interest in using furfural and LA as a feedstock for biofuels and chemicals is fast increasing as clearly indicated from the number of publications and citations on this subject in the past five years (Figure 2.3).

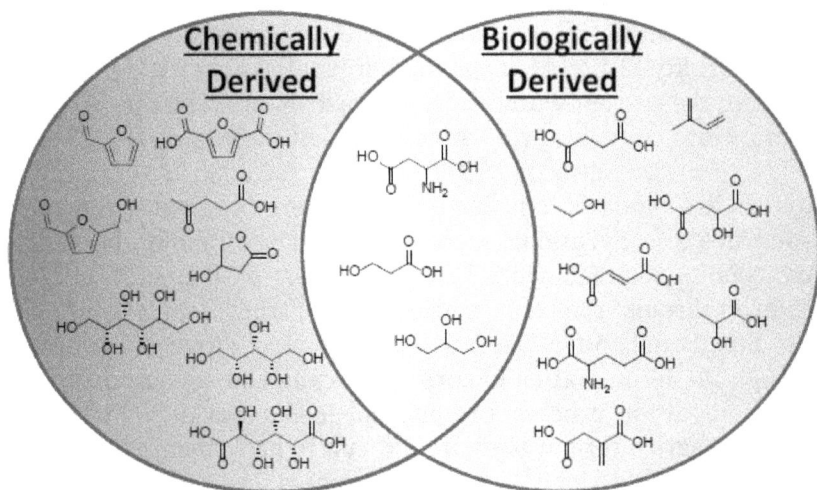

Figure 2.1 Top bio-renewable chemicals produced via chemical processes, biological processes, or both (from the DOE "Top 10" list [32] and its update [33]).

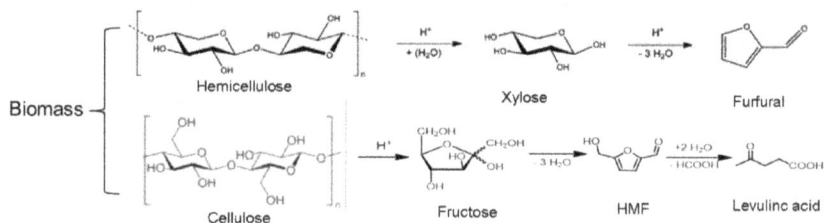

Figure 2.2 Schematic for the transformation of biomass to furfural and levulinic acid.

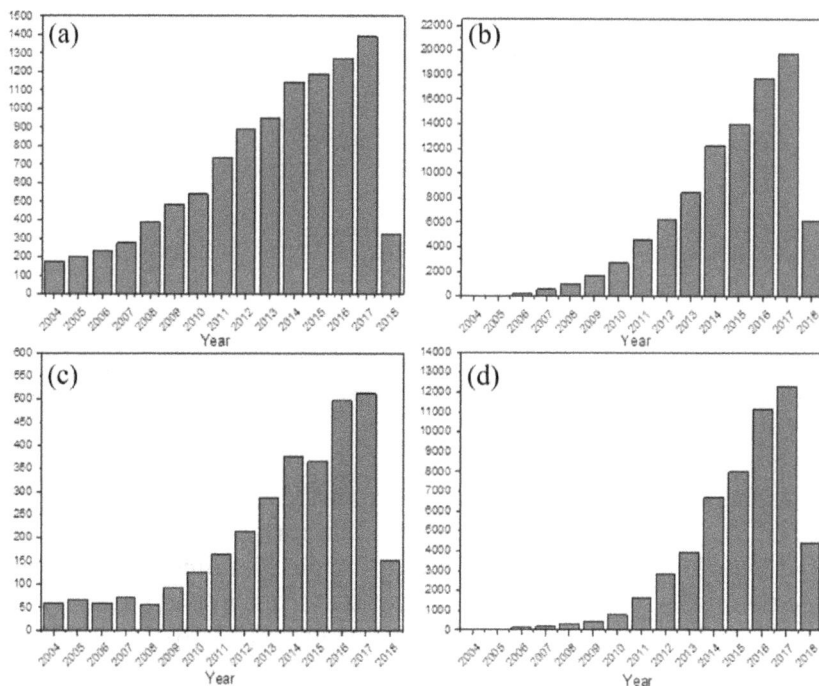

Figure 2.3 Number of publications and citations on furfural (a and b) as well as levulinic acid (c and d) from January 2004 to April 2018.

Many highly value-added chemicals (e.g., furfuryl alcohol, furanones, furans, 2-furoyl chloride) can be produced from furfural which has a global production of ~300 kton/y [38-40]. According to the market analysis, North America, Europe, Asia Pacific, the Middle East and Africa are the main consumers of furfural. In 2014, Asia

Pacific contributed 78% to the overall demand of furfural. It is estimated to rise by 4.5% by 2023, possibly because of furfural-derived
agrochemicals, herbicides and pesticides. Besides, furfural provides
numerous opportunities to generate fuels and valuable chemicals as
shown in Figure 2.4 [36].

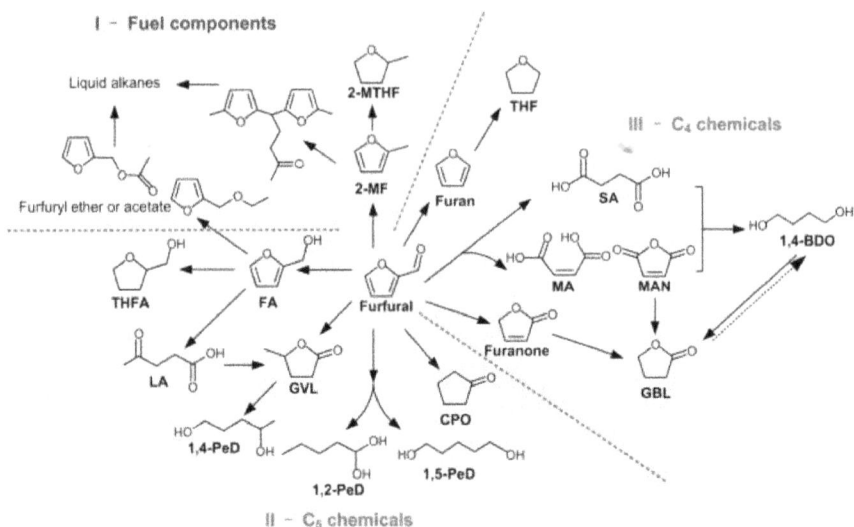

Figure 2.4 Transformation of furfural to valuable chemicals and fuels.
Reproduced from Reference 41 with permission from American Chemical
Society.

As mentioned earlier, DOE/NREL report [32] and the update from
Bozell *et al.* [33] presented the crucial importance of the bio-based
platform chemical levulinic acid (LA) for the biorefinery. Numerous
fine chemicals and fuels can be synthesized from levulinic acid, which
is considered as a substitute for fossil resources [42-48]. For instance,
it is a crucial intermediate for useful fine chemicals such as polymer
resins, textile dyes, fuel additives, herbicides and plasticizers [27,49-
51].

Levulinic acid is generally obtained through multi-step hydrolysis
of native biomass via 5-hydroxymethylfurfurfural (HMF) intermediate, as depicted in Figure 2.2. The native biomass resources are also
often subjected to pre-treatment and detoxification [52,53]. Due to
the presence of keto and carboxylic functional groups, a wide range
of derivatives, depicted in Figure 2.5, can be selectively produced
from levulinic acid [16,44,54].

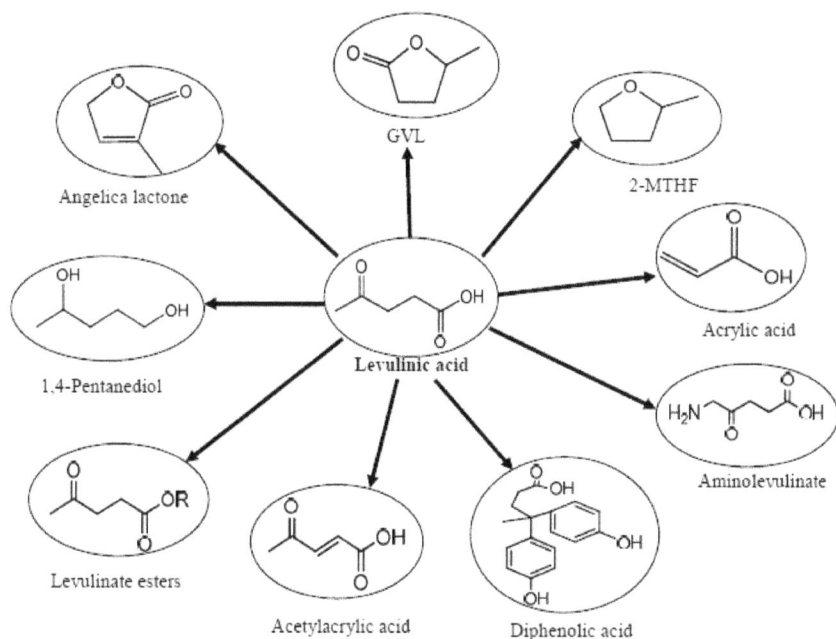

Figure 2.5 Schematic of the fine chemicals and fuels additives synthesized from LA feedstock [16,27]. Reproduced from Reference 27 with permission from Elsevier.

2.2 Catalytic Upgrading of Furfural to Classic Fuels

Furfural, generally obtained from dehydration of xylose, has the potential to be converted to fuels or fuel additives and is getting increasing attention for attaining widespread application. Furfural itself is not an ideal fuel component with respect to stability and melting point. Nevertheless, it can act as a platform to synthesize a large number of fuel additive and high-value chemicals. As a building block with great promise, it represents a rich resource to generate 2-methylfuran (MF), 2-methyltetrahydrofuran (MTHF), tetrahydrofurfuryl alcohol (THFA) and tetrahydrofuran (THF). Figure 2.6 also shows different strategies for converting furfural to biofuels.

2.2.1 Synthesis of MF and MTHF

It is unambiguous that low oxygen content as well as high carbon or

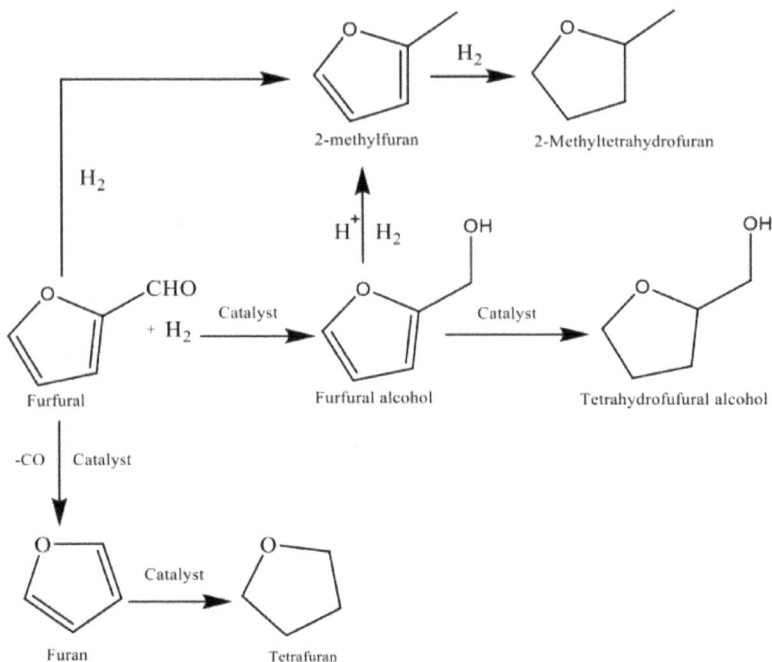

Figure 2.6 Typical routes for converting furfural to fuel additive and high value chemicals. Reproduced from Reference 39 with permission from Elsevier.

hydrogen content is an indispensable property of desired fuels. However, furfural molecule has high oxygen content, thus, the hydrogenation of furfural is an effective route to lower oxygen content and enlarge hydrogen content. The hydrogenation of the carbonyl to methyl group leads to the formation of MF and further deep hydrogenation to MTHF. MF and MTHF have been reported as highly promising biofuel components that can be mixed with gasoline used as transportation fuel. Representative physical properties are listed in Table 2.1 to compare with biofuel ethanol and gamma-valerolactone (GVL).

The synthesis of MF from furfural (Figure 2.7) has been investigated over the noble metal and bimetallic catalysts via vapor phase hydrogenation. The furfuryl alcohol (FA) intermediate was verified by GC and GC-MS in relevant literature studies [40,58-61]. Various kinds of Cu based catalysts exhibited high selectivity for the production of MF in spite of the disadvantage of swift deactivation of catalyst by coking, which is often performed under high temperature and low pressure for further step of the hydrogenolysis of FA to form MF, thus,

Table 2.1 Representative physical properties of the potential fuel components [55-57]

Terms	ethanol	MF	GVL	MTHF
Mass Weight $(g \cdot mol^{-1})$	46.07	82.10	100.12	86.13
Carbon (wt%)	52.2	73.15	60	69.72
Hydrogen (wt%)	13.1	7.37	8	11.70
Oxygen (wt%)	34.7	19.49	32	18.58
Boiling point (°C)	78	64	207	80.3
Melting point (°C)	-114	-89	-31	-136
Flash point (°C)	13	-22	96	-12
Density $(g \cdot ml^{-1})$	0.789	0.91	1.046	0.854
Solubility in water (mg/ml)	miscible	3	>=1	150
Enthalpy of vaporization (kJ/kg)	912	357	442.36	364.43

leading to the rapid deactivation of the metal active center. Considering the drawbacks, the catalyst was regenerated by coke burn off at 400 °C [62,63]. Multi-metal Cu based catalysts and carbon-supported Cu based catalysts were confirmed to be efficient in the hydrogenation of furfural to generate MF, however, deactivation of catalysts is an inescapable factor limiting its wide commercial application as commercial operations require highly stable catalysts and efficient reaction protocols. To benchmark different catalysts used for the hydrogenation/hydrolysis of furfural to MF, some classical works and system from previous literature studies have been summarized in Table 2.2. The ongoing research in this direction has been focused to understand the interior relations between different variables and selectivity of MF to further illuminate key mechanistic issues between the variables and associated performance [64].

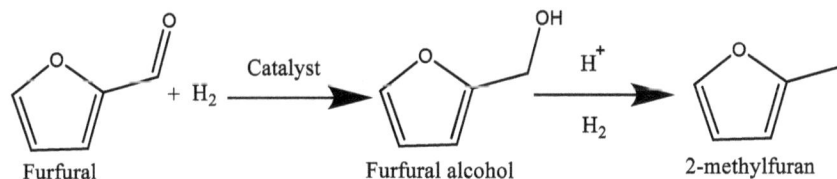

Figure 2.7 Hydrogenation of furfural to 2-methylfuran.

The efficient production of MTHF (Figure 2.8) can be obtained by two procedures, which include the hydrogenation of furfural and LA. The valorization of furfural to form MTHF mainly includes two steps:

Table 2.2 Typical studies on the hydrogenation of furfural to MF employing heterogeneous catalysts. Updated and partially reproduced from Reference 39 with permission from Elsevier.

No.	Catalyst	Reaction conditions	Conv. (%)	Y_{MF} (%)	Ref.
1	5% Pt/C	175 °C, 80 bar H_2,0.5 h, n-butanol solvent	99.3	40.4	[65]
2	5% Pt/C	150 °C, 20 bar H_2, 4 h, acetic acid-assisted	41.2	8.9	[66]
3	5% Pt/C	175 °C, 30 bar H_2, 1 h, 20 ml, H_2O solvent H_3PO_4(85%)	100	36.6	[67]
4	Cu-Zn-Al	225 °C, 6 h, LHSV=0.7 h^{-1}	99.9	93.0	[68]
5	2 wt% Pt/TiO_2/SiO_2	150 °C, HLSV=2 h^{-1}, H_2/furfural=2 mol/mol)	69.4	18.9	[69]
6	Cu-Mn-Si	279 °C, 8 h, 1 atm, LHSV=0.49 h^{-1}, n(H_2):n(CHL+FFA)=1 0:1, coupling reaction	99.8	93.5	[70]
7	Cu/Zn/Al/Ca/Na=59:33 :6:1:1	250 °C, LHSV=0.3 h^{-1}, H_2:furfural =25 (molar ratio)	99.7	87	[71]
8	CuO/$CuFe_2O_4$	220 °C,90 bar H_2, 14 h	99.4	51.1	[72]
9	CuLa-β zeolite	180 °C, 1 bar, H_2/fur-fural=5, CHSV 0.087 mol h^{-1} g^{-1}_{catal}	9.5	7.8	[73]
10	Cu-Zn-Al oxide	150 °C, 1 bar, H_2/fur-fural=10, LHSV 0.087 mol h^{-1} g^{-1}_{catal}	99	86.1	[74]
11	Zincian malachite Cu/ZnO	200 °C, ambient pres-sure, 24 h, LHSV=0.5 h^{-1}	99.7	76.9	[75]
12	Cu/AC-400/2	170 °C, 4 MPa H_2, 4 h, 2-propanol as solvent	100	100	[76]
13	Cu-Co/Al_2O_3	493 K, 4 MPa H_2, 4 h	100	78	[77]
14	10%Ni-10%Cu/Al_2O_3	210 °C, formic acid as hydrogen donor, 4 h, isopropanol solvent	97.4	75.6	[78]
15	Ru_4/$NiFe_2O$	180 °C, 2-propanol as the hydrogen source, 2.1 MPa N_2, 4 h	98.4	68.7	[79]
16	Pd/C	220 °C, 500 psi H_2, Isopropyl alcohol sol-vent, 5 h	>99	18	[80]
17	10Cu-1Pd/ZrO_2	220 °C, PrOH hydro-gen donor, 4 h	98.5	63.6	[81]

In the first step, MF is formed via hydrogenation of carbonyl group and further hydrogenolysis of FA to MF. In the second step, the furan ring of MF is hydrogenated to form MTHF. Diverse Ru-derived catalysts, noble-metal catalysts, Cu-composites and other catalysts have been designed and utilized for the production of MTHF (Table 2.3). It is worth mentioning that a two-step conversion of furfural to produce MTHF was also proposed [82]. Cu chromite and Pd/C were employed under supercritical CO_2,. The advantage of the process is that FA, THFA, MF, MTHF or furan can be obtained by adjusting the reaction temperatures. Besides, under hasher conditions, noble-metal catalysts can lead to an efficient conversion of the furan ring to produce MTHF.

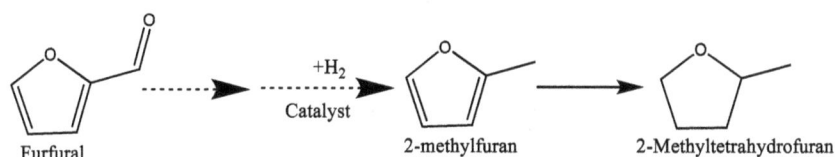

Furfural   2-methylfuran   2-Methyltetrahydrofuran

Figure 2.8 Catalytic conversion of furfural to MTHF.

2.2.2 THFA

Tetrahydrofurfuryl alcohol (THFA) is eco-friendly and has attractive properties including mobility, transparency, high-boiling point and miscibility with water. It has many practical applications such as agricultural needs and printing inks as well as industrial and electronics cleaners. On average, the Japanese industry annually synthesizes ~30 tons THFA commercially. On the laboratory scale, THFA is usually synthesized in the two-step conversion of furfural. In the conversion process of furfural (Figure 2.9), furfural alcohol serves as intermediate during the hydrogenation reaction catalyzed by CuCr-derived from hydrotalcite and transitional noble metal catalysts (e.g., Pd, Au), respectively [59,61,82,86]. The homogeneous Ru(II) bis(diimine) catalyst displays ~26% THFA selectivity from the catalytic conversion of furfural [87]. On the whole, Ni-based catalysts exhibit better selectivity for both hydrogenation of furfural or furfural alcohol. On the industrial level, Ni-based catalysts are often employed for converting furfural alcohol to THFA, where both vapor or liquid phase are economically feasible at temperature in the range 50 to 100 °C. In

Table 2.3 Typical studies on the hydrogenation of furfural to MTHF employing different catalysts. Partially reproduced from Reference 39 with permission from Elsevier.

No.	Catalyst	Reaction conditions	Conv. (%)	Y_{HTHF} (%)	Ref.
1	NiCu/SBA-15	160 °C, 40 bar H_2, 4 h, water solvent	>99	>16.8	[83]
2	NiCu/SBA-15	160 °C, 40 bar H_2, 4 h, water solvent, Na_2HPO_4 additive	>99	>35.6	[83]
3	NiCu/SBA-15	160 °C, 40 bar H_2, 4 h, water solvent, Na_2CO_3 additive	>99	>27.7	[83]
4	3% Pd/C	160 °C, 80 bar H_2, 0.5 h, water solvent	99.8	16.7	[67]
5	5% Pt/C	175 °C, 80 bar , 0.5 h, water solvent	100	9.4	[84]
6	Rh-ReOx/SiO₂	50 °C+120 °C, 2 h+24 h, 60 bar H_2, water solvent	>99.9	26.9	[85]
7	Pd-Ir-ReOx/SiO₂	120 °C, 24 h, 60 bar H_2, water solvent	>99.9	11.9	[85]
8	Pd-Ir-ReOx/SiO₂	40 °C+120 °C, 2 h+24 h, 20 bar H_2, water solvent	>99.9	12.5	[85]
9	5% Pd/C	175 °C, 80 bar H_2, 1 h, water solvent	100	35.6	[65]
10	5% Ru/C	175 °C, 80 bar H_2, 1 h, water solvent	100	6.1	[65]
11	CoMnCr	175 °C, 80 bar H_2, 0.5 h, water solvent	100	16.5	[65]
12	Raney Ni Actimet C	160 °C, 30 bar H_2, 1 h, water solvent	100	23.4	[65]
13	4%Pd/C	220 °C, 500 psi H_2, Isopropyl alcohol solvent, 5 h	>99	40	[80]
14	5%Pd/C	220 °C, 500 psi H_2, Isopropyl alcohol solvent, 5 h	>99	35	[80]
15	4%Ru/C	220 °C,500 psi H_2, Isopropyl alcohol solvent, 5 h	>99	37	[80]
16	Cu-Pd/ZrO	220 °C, PrOH hydrogen donor, 4 h	100	78.8	[81]
17	Cu/ZrO₂+Pd / ZrO₂	220 °C, PrOH hydrogen donor, 4 h	97.9	48.3	[81]
18	Pd/ZrO₂	220 °C, PrOH hydrogen donor, 4 h	97.2	33.5	[81]

a significant study, Rode *et al.* [88] also reported full selectivity of 95% THFA with a near-perfect conversion of furfural in one pot over a Pd impregnated in Si-MFI molecular sieve catalyst (3% Pd/MFI). Table 2.4 summarizes representative studies on the production of THFA via furfural conversion. In short, effective THFA selectivity is possible if the metal catalysts can be modified and the deactivation can be reduced.

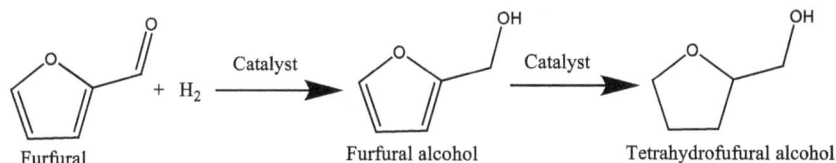

Figure 2.9 Catalytic conversion of furfural to THFA.

2.2.3 Tetrahydrofuran (THF)

Hydrogenation of furan to produce tetrahydrofuran (THF) is one of the most appealing strategies for furan application. THF is widely used as a solvent for organic reagents for preparing complex catalysts or chromatographic techniques. Besides, THF is also used in many applications such as vinyl films, PVC cement, adhesives and cellophane [99,100]. On the laboratory scale, THF is often produced from the hydrogenation of furan (Figure 2.10). Producing THF from furan is obviously promising as furan is obtained from furfural, which in turn is originally acquired from biomass, like hemicellulose. The reaction proceeds in the presence of noble metals and Ni-based catalysts [101,102]. During the reaction, high yield of coke formation could lead to the deactivation of catalysts and restrain the formation of THF. On commercial level, THF is produced via the catalytic hydrogenation of maleic anhydride, patented by Du Pont [103].

2.3 Conversion of Levulinic Acid to Typical Fuels

2.3.1 γ-Valerolactone

Over the last few decades, research studies have reported the beneficial properties of γ-valerolactone, such as low melting point (-31 °C), high boiling point (207 °C), flash point (96 °C), a definite but suitable

Table 2.4 Typical studies on the hydrogenation of furfural to THFA. Partially reproduced from Reference 39 with permission from Elsevier.

No.	Catalyst	Reaction conditions	Conv. (%)	Y_{THFA} (%)	Ref.
1	NiO/SiO$_2$	200 °C, 1 bar	15	>14.9	[89]
2	Ni/SiO$_2$	140 °C, 1 bar H$_2$, GHSV=1.1 mol h^{-1} g catalyst^{-1}, N$_2$ protecting	>99	>93.1	[90]
3	Raney Ni/Al(OH)$_3$	110 °C, 30 bar H$_2$, 1.25 h, isopropanol solvent	>99	>99	[91]
4	Ni-Pd/SiO$_2$	40 °C, 80 bar, 8 h, water solvent	99	95.0	[92]
5	RuO$_2$	120 °C, 50 bar H$_2$, 2.5 h, methanol solvent	100	~76	[93]
6	Ni 5132P+Cu V1283	130 °C, 40 bar H$_2$, 3 h 5 min, methanol solvent	100	97	[93]
7	Ni 473P+Cu V1283	130 °C, 40 bar H$_2$, 3 h 10 min, methanol solvent	99	95	[93]
8	RuO$_2$+Cu V1283	120 °C, 50 bar H$_2$, 1 h 35 min, methanol solvent	100	86	[93]
9	Pd/C+Cu V1283	120 °C, 50 bar H$_2$, 3.5 h , methanol solvent	99	28	[93]
10	Ru/C	165 °C, 25 bar H$_2$, 1-Butanol-water solvent	100	16.6	[94]
11	Ru/C	165 °C, 25 bar H$_2$, MTHF solvent	91	11.2	[94]
12	Ni-Sn	Iso-PrOH solvent, 30 bar H$_2$, 110 °C, 1.25 h	16	4	[91]
13	Ru$_4$/NiFe$_2$O	180 °C, 2-propanol as the hydrogen source, 2.1 MPa N$_2$, 4 h	98.4	2.0	[79]
14	Ru/C	220 °C, 500 psi H$_2$, Isopropyl alcohol solvent, 5 h	>99	25	[80]
15	NiCoB/H$^+$-ATP-A	100 °C, 3 MPa H$_2$, ethanol, 2 h	81.9	17.5	[95]
16	(10) W$_x$C-β-SiC	90 °C, 20 bar H$_2$, 2 h	100	90.5	[96]
17	(20) W$_x$C-β-SiC	90 °C, 20 bar H$_2$, 2 h	67.6	24.6	[96]
18	10Cu-3Pd/SiO$_2$	220 °C, PrOH, 4 h	100	22.6	[81]
19	Pd@MIL-101(Cr)-NH$_2$	40 °C, 2 MPa H$_2$, water, 6 h	>99.9	>99.9	[97]
20	Pd-HAP	40 °C, 2-propanol, 1 MPa H$_2$, 4 h	100	100	[98]

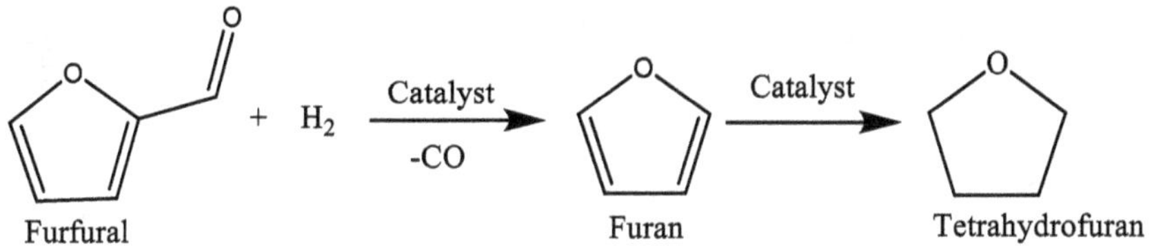

Figure 2.10 Catalytic conversion of furfural to THF.

smell for easy recognition of leaks and spills, low toxicity and high solubility in water to assist biodegradation [11,104]. As shown in Table 2.5, GVL can be mixed with gasoline, where it exhibits similar fuel properties as ethanol/gasoline mixture [27,104,105]. Besides, no peroxides is formed in air for weeks, making it a safe material for large scale use [106]. The group of Horváth studied that the use of GVL as a sustainable liquid fuel enabled its worldwide monitoring and regulation [57,107,108]. The attractive benefit of GVL to be a practical biofuel is its low cost due to its synthesis from renewable biomass.

Table 2.5 Physical properties of potential biofuels. Reproduced from Reference 16 with permission from Elsevier.

Terms	Ethanol	GVL	MTHF	EL
M (g mol^{-1})	46.07	100.12	86.13	144.17
Carbon (wt %)	52.2	60	69.7	58.7
Hydrogen (wt %)	13.1	8	11.6	7.7
Oxygen (wt %)	34.7	32	18.7	33.5
Boiling point (°C)	78	207	80	206.2
Melting point (°C)	-114	-31	-136	---
Flash Point (°C)	13	96.1	-11.1	195
Density (g mL^{-1})	0.789	1.0485	0.86	1.014
Solubility in water (mg mL^{-1})	Miscible	≥ 100	13	soluble
Octane number [a]	108.6	---	80	---
Cetane number	5	---	23.5[c]	<10
Lubricity	---	---	---	287

a: research octane number

Over the last a few decades, a number of studies have tried to selectively produce GVL using advanced systems (Figure 2.11). Homogeneous catalysts have exhibited high selectivity for the synthesis of GVL through LA hydrogenation [107,109]. Solid catalysts offer great opportunities and environmentally benign properties, especially for the separation of catalyst from reaction residues. For instance, Ru- and Cu-derived nanoparticle based catalysts have been reported to be effective for the selective production of GVL [45,72, 105,110]. Our previous works also documented that Cu-catalysts derived from hydrotalcite and Pd-nanoparticles were highly selective for the synthesis of GVL [45,72,105,110,111].

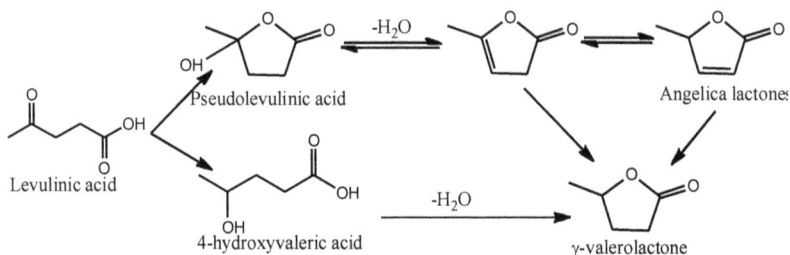

Figure 2.11 The classic pathway for the formation of GVL from LA. Reproduced from Reference 16 with permission from Elsevier.

Table 2.6 also presents a summary of the recent developments about the conversion of LA to GVL. Vapor phase hydrogenation of LA using H_2 is often performed at high temperature, associated with high conversion and high yield of GVL. However, it requires high energy input and special facility. Liquid phase hydrogenation of LA is generally performed at low temperature. However, the liquid phase often exhibits low conversion and relatively low selectivity of GVL. In-situ hydrogenation of LA to GVL without external H_2 source is more promising, especially when the raw biomass monomer is used.

2.3.2 2-Methyltetrahydrofuran

2-Methyltetrahydrofuran (MTHF) is not miscible with water and its solubility decreases with increasing temperature [39,123]. MTHF is a highly flammable mobile liquid with a melting point of -136 °C and boiling point of 80.2 °C. GVL faces the issues of high water solubility, possible corrosiveness in storage and lower energy density, thus, upgrading of GVL to MTHF is more attractive approach. It has been reported that MTHF can be added up to 70% in gasoline. Important properties of MTHF have been compared with other fuels in Table 2.5. Figure 2.12 also demonstrates the catalytic transformation of LA to MTHF.

Table 2.6 Hydrogenation of LA to GVL using various catalysts. Reproduced from Reference 16 with permission from Elsevier.

No.	Sub-strate	Catalyst	Solvent	T (°C)	H_2 sources	t (h)	Y_{GVL} (%)	Ref.
1	LA	Ru/C	Dioxane	150	34.5 bars H_2	4	97	[112]
2	LA[a]	Ru/SiO$_2$	H$_2$O+CO$_2$	200	100 bars H_2	N.D.	>99	[113]
3	LA	Ru/TiO$_2$	H$_2$O	150	HCOOH +40 bars H_2	1	63	[114]
4	LA	Ru/C	CH$_3$OH	130	12 bars H_2	3	91	[115]
5	LA	Pd/Al$_2$O$_3$	------	220	HCOOH	12	29	[116]
6	LA	5% Ru/C	Dioxane	265	1-25 bars H_2	50	98.6	[117]
7	LA	5% Rh/C	Dioxane	141	H_2+CO$_2$ (247 bars)	N.D.	98.9	[117]
8	LA	Au/ZrO$_2$-VS	H$_2$O	180	Formic acid	3	99	[118]
9	LA	Au/ZrO$_2$	H$_2$O	170	BF	6	95	[119]
10	BL[b]	5%Au/ZrO$_2$	H$_2$O	150	Formic acid	5	97	[120]
11	Fruc-tose	5%Au/ZrO$_2$	H$_2$O	140	Formic acid	5	66	[120]
12	Fruc-tose	Ru–Ni/ Meso-C	---	150	45 bars H_2	2	96	[121]
13	LA	5% Ru/ Hydroxyap-atite	H$_2$O	70	5 bars H_2	4	99	[122]

Note: Y: yield of GVL; a, N.D.: not defined; b, BL: butyl levulinate.

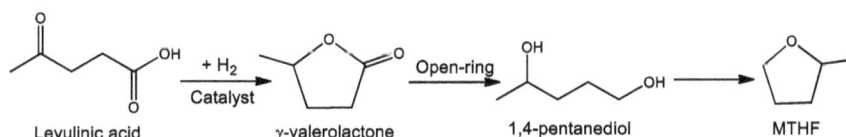

Figure 2.12 Catalytic transformation of LA to MTHF.

Based on previous literature studies, two pathways (Figure 2.13) for MTHF synthesis have been well studied: LA dehydration via an-gelicalactone (AGL) and LA hydrogenation to 4-hydroxypentanoic acid (HPA). Recently, Al-Shaal *et al.* [123] confirmed that AGL was easily converted into MTHF using a commercial Ru/C catalyst under solvent-free conditions. The other hydrogenation path in water medium often leads to the formation of HPA intermediates. The reaction products can be tuned using different catalysts and experimental conditions. The transformation of LA into GVL, 1,4-PDO, and MTHF involves consecutive series of hydrogenation/dehydration paths as

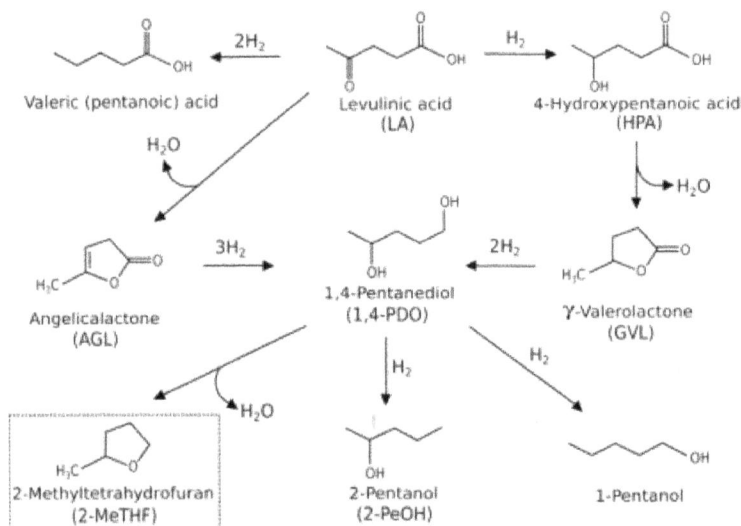

Figure 2.13 Two pathways for the synthesis of MTHF from LA. Reproduced from Reference 27 with permission from Elsevier.

shown in Figure 2.14 [109,124]. So far, the biggest challenge for the heterogeneous conversion of LA to MTHF is the large amount of coke

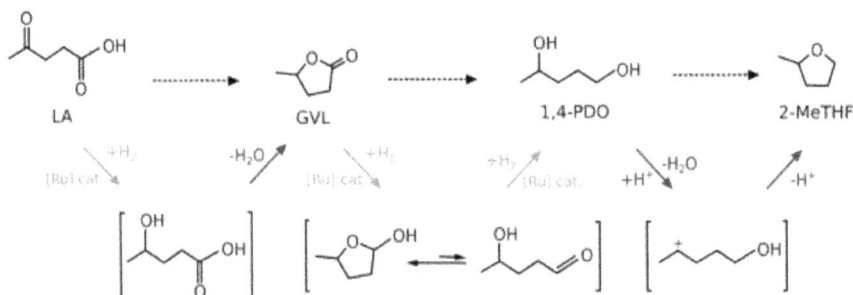

Figure 2.14 The mechanism for the formation of MTHF. Reproduced from Reference 109 with permission from American Chemical Society.

formation. It has been observed that the addition of hydrogen donor or hydrogen could reduce/hinder coke formation to some extent. Besides, the low-cost purification of MTHF still requires further efforts.

2.3.3 Levulinate Esters

Along with the value-added derivatives from the transformation of LA, alkyl levulinates (Figure 2.15) are attractive because of their wide range of applications in flavoring, solvent and plasticizer sectors [20]. These levulinate esters, like methyl levulinate, ethyl levulinate and butyl levulinate, are short chain fatty esters exhibiting low toxicity, high lubricity, stable flash point and moderate flow properties, thus, making alkyl levulinates suitable additives for gasoline and diesel fuels [125]. On using methanol and ethanol, the high yields of levulinate esters (e.g., methyl levulintae and ethyl levulinate) can be obtained at low production cost [33,126-128].

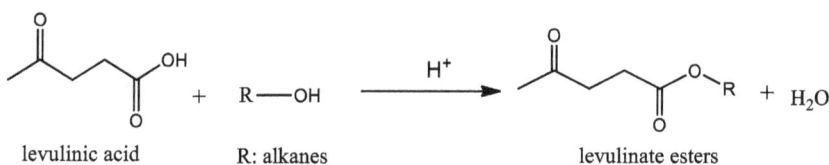

levulinic acid   R: alkanes   levulinate esters

Figure 2.15 Synthesis of levulinate esters from LA and alcohols.

Over the last few decades, it has been observed that homogeneous catalysts (e.g., formic acid, H_3PO_4 and H_2SO_4) are efficient for the esterification of LA with alcohols. However, the homogeneous catalyst face the issues such as catalyst reuse, product separation and environmental protection. The heterogeneous catalysts like supported heteropolyacid [127,129,130], zeolites [131], sulfonated metal oxides [132] and silicas [6], sulfonated carbon nanotubes [133], Starbon® mesoporous carbon [134] and hybrid catalysts [135] are also useful.

2.4 Future Outlook

Overall, a brief review on the catalytic upgrading of furfural and levulinic acid for the production of alternative fuels has been presented. A rapid acceleration of studies has concentrated on developing highly efficient catalysts as well the reaction systems for the effective conversion of furfural and levulinic acid. However, the development of cost-effective processes avoiding the separation of furfural and levulinic acid from reaction system still face a lot of challenges. One-step conversion of biomass to fuels without complex pretreatment and isolation of intermediates will be highly desirable.

Mineral acid catalysts are used widely at the lab-scale, however, for the industrial applications, separation of the products with high purity, recycling of catalyst and safe handling still face a lot of challenges. Heterogeneous acidic catalysts are easier to be recovered and recycled, however, these often display lower performance in terms of conversion and yield. Occasionally, the utilized catalysts have to be regenerated after each run. The development of non-toxic, low-cost and easily recyclable catalysts is a promising approach. The studies dealing with the process parameters to enhance separation and final yield still need to develop more in-depth understanding. Specifically, such understanding can be gained by studying catalytic systems, catalyst properties controlling bond scission, role of secondary metals in the reactive chemistry for catalysts, influence of molecular functionality on reaction pathways and kinetic requirements needed to maximize activity and selectivity. The use of operando tools would also provide further insights into the reaction processes, especially for catalyst deactivation. Furthermore, the development of biphasic systems combined with solid catalysts containing acid functionality can also be interesting for future studies.

Acknowledgments

The work was supported by National Natural Science Foundation of China (21776324), National Key R&D Program of China (2018YFD0800700) and "Hundred Talent Plan" from Sun Yat-sen University.

Abbreviations

MF: 2-methylfuran
MTHF: 2-methyltetrahydrofuran
THFA: tetrahydrofurfuryl alcohol
THF: tetrahydrofuran
DOE: Department of Energy
NREL: National Renewable Energy Laboratory
HMF: 5-hydroxymethylfurfurfural

GVL: gamma-valerolactone
FA: furfuryl alcohol
LHSV: liquid hourly space velocity
WHSV: weight hourly space velocity
LA: levulinic acid
ML: methyl levulinate
EL/BL: ethyl/butyl levulinate

References

1. Huber, G. W., Iborra, S., and Corma, A. (2006) Synthesis of transportation fuels from biomass: chemistry, catalysts, and engineering. *Chemical Reviews*, **106**, 4044-4098.
2. Alonso, D. M., Bond, J. Q., and Dumesic, J. A. (2010) Catalytic conversion of biomass to biofuels. *Green Chemistry*, **12**, 1493-1513.
3. Luque, R., De, S., and Balu, A. (2016) Catalytic conversion of biomass. *Catalysts*, **6**, 148.
4. Yan, K., and Lu, Y. (2016) Direct growth of MoS$_2$ microspheres on Ni Foam as a hybrid nanocomposite efficient for oxygen evolution reaction. *Small*, **22**, 2975-2981.
5. Yan, K., Maark, T. A., Khorshidi, A., Sethuraman, V. A., Peterson, A. A., and Guduru, P. R. (2016) The influence of elastic strain on catalytic activity in the hydrogen evolution reaction. *Angewandte Chemie, International Edition*, **55**, 6175-6181.
6. Giang, C., Osatiashtiani, A., dos Santos, V., Lee, A., Wilson, D., Waldron, K., and Wilson, K. (2014) Valorisation of Vietnamese rice straw waste: Catalytic aqueous phase reforming of hydrolysate from steam explosion to platform chemicals. *Catalysts*, **4**, 414.
7. Yan, K., Lafleur, T., Chai, J., and Jarvis, C. (2016) Facile synthesis of thin NiFe-layered double hydroxides nanosheets efficient for oxygen evolution. *Electrochemistry Communications*, **62**, 24-28.
8. Yan, K., Lu, Y., and Jin, W. (2016) Facile synthesis of mesoporous manganese-iron nanorod arrays efficient for water oxidation. *ACS Sustainable Chemistry & Engineering*, **4**, 5398-5403.
9. Antonetti, C., Licursi, D., Fulignati, S., Valentini, G., and Galletti, A. R. (2016) New frontiers in the catalytic synthesis of levulinic acid: from sugars to raw and waste biomass as starting feedstock. *Catalysts*, **6**, 196.
10. Weng, Y., Qiu, S., Ma, L., Liu, Q., Ding, M., Zhang, Q., Zhang, Q., and Wang, T. (2015) Jet-fuel range hydrocarbons from biomass-derived sorbitol over Ni-HZSM-5/SBA-15 catalyst. *Catalysts*, **5**, 2147.
11. Yan, K., Lafleur, T., Wu, X., Chai, J., Wu, G., and Xie, X. (2015) Cascade upgrading of [gamma]-valerolactone to biofuels. *Chemical Commun-*

ications, **51**, 6984-6987.

12. Al Maksoud, W., Larabi, C., Garron, A., Szeto, K. C., Walter, J. J., and Santini, C. C. (2014) Direct thermocatalytic transformation of pine wood into low oxygenated biofuel. *Green Chemistry*, **16**, 3031-3038.

13. Alonso, D. M., Wettstein, S. G., and Dumesic, J. A. (2012) Bimetallic catalysts for upgrading of biomass to fuels and chemicals. *Chemical Society Reviews*, **41**, 8075-8098.

14. An, H. J., Wilhelm, W. E., and Searcy, S. W. (2011) Biofuel and petroleum-based fuel supply chain research: A literature review. *Biomass & Bioenergy*, **35**, 3763-3774.

15. Yan, K., Wu, G., and Jin, W. (2016) Recent advances in the synthesis of layered, double-hydroxide-based materials and their applications in hydrogen and oxygen evolution. *Energy Technology*, **4**, 354-368.

16. Yan, K., Yang, Y., Chai, J., and Lu, Y. (2015) Catalytic reactions of gamma-valerolactone: A platform to fuels and value-added chemicals. *Applied Catalysis B: Environmental*, **179**, 292-304.

17. Abideen, Z., Hameed, A., Koyro, H. W., Gul, B., Ansari, R., and Khan, M. A. (2014) Sustainable biofuel production from non-food sources - An overview. *Emirates Journal of Food & Agriculture*, **26**, 1057-1066.

18. Corma, A., Huber, G. W., Sauvanaud, L., and O'Connor, P. (2007) Processing biomass-derived oxygenates in the oil refinery: Catalytic cracking (FCC) reaction pathways and role of catalyst. *Journal of Catalysis*, **247**, 307-327.

19. Hromadko, J., Miler, P., Honig, V., and Cindr, M. (2010) Technologies in second-generation biofuel production. *Chemicke Listy*, **104**(8), 784-790.

20. Corma, A., Iborra, S., and Velty, A. (2007) Chemical routes for the transformation of biomass into chemicals. *Chemical Reviews*, **107**, 2411-2502.

21. Lin, Y.-C., and Huber, G.W. The critical role of heterogeneous catalysis in lignocellulosic biomass conversion. (2009) *Energy & Environmental Science*, **2**, 68-80.

22. Carlson, T., Tompsett, G., Conner, W., and Huber, G. (2009) Aromatic production from catalytic fast pyrolysis of biomass-derived feedstocks. *Topics in Catalysis*, **52**, 241-252.

23. Song, Y., Wang, X., Qu, Y., Huang, C., Li, Y., and Chen, B. (2016) Efficient dehydration of fructose to 5-hydroxy-methylfurfural catalyzed by heteropolyacid salts. *Catalysts*, **6**, 49.

24. Yan, K., Wu, X., An, X., and Xie, X. (2014) Facile synthesis of reusable coal-hydrotalcite catalyst for dehydration of biomass-derived fructose into platform chemical 5-hydroxymethylfurfural. *Chemical Engineering Communications*, **201**, 456-465.

25. Yan, K., and Luo, H. (2015) Recent development of metal nanoparti-

cles catalysts and their use for efficient hydrogenation of biomass-derived levulinic acid. In: *Green Processes for Nanotechnology*, Basiuk, V. A., and Basiuk, E. V. (eds.), Springer, Switzerland, pp. 75-98.

26. Climent, M. J., Corma, A., and Iborra, S. (2010) Heterogeneous catalysts for the one-pot synthesis of chemicals and fine chemicals. *Chemical Reviews*, **111**, 1072-1133.

27. Yan, K., Jarvis, C., Gu, J., and Yan, Y. (2015) Production and catalytic transformation of levulinic acid: A platform for specialty chemicals and fuels. *Renewable* and *Sustainable Energy Reviews*, **51**, 986-997.

28. Chakraborty, S., Aggarwal, V., Mukherjee, D., and Andras, K. (2012) Biomass to biofuel: A review on production technology. *Asia-Pacific Journal of Chemical Engineering*, **7**, S254-S262.

29. Dutta, S., De, S., Saha, B., and Alam, M. I. (2012) Advances in conversion of hemicellulosic biomass to furfural and upgrading to biofuels. *Catalysis Science & Technology*, **2**, 2025-2036.

30. Zhang, Y. H. P., Sun, J. B., and Zhong, J. J. (2010) Biofuel production by in vitro synthetic enzymatic pathway biotransformation. *Current Opinion in Biotechnology*, **21**, 663-669.

31. Zinoviev, S., Muller-Langer, F., Das, P., Bertero, N., Fornasiero, P., Kaltschmitt, M., Centi, G., and Miertus, S. (2010) Next-generation biofuels: survey of emerging technologies and sustainability issues. *ChemSusChem*, **3**, 1106-1133.

32. Werpy, T., Petersen, G., Holladay, J., White, J., and Manheim, A. (2004) Top Value Added Chemicals from Biomass, Volume I - Results of Screening for Potential Candidates from Sugars and Synthesis Gas. *Report of National Renewable Energy Laboratory*, USA. Online: https://www.nrel.gov/docs/fy04osti/35523.pdf (assessed 19th June 2018).

33. Bozell, J. J., and Petersen, G. R. (2010) Technology development for the production of biobased products from biorefinery carbohydrates - the US Department of Energy's "Top 10" revisited. *Green Chemistry*, **12**, 539-554.

34. Gallezot, P. (2012) Conversion of biomass to selected chemical products. *Chemical Society Reviews*, **41**(4), 1538-1558.

35. Adler, P. R., Sanderson, M. A., Boateng, A. A., Weimer, P. J., and Jung, H. J. G. (2006) Biomass yield and biofuel quality of switchgrass harvested in fall or spring. *Agronomy Journal*, **98**, 1518-1525.

36. Yan, K., and Luo, H. (2017) Production of γ-Valerolactone from Biomass. In: *Production of Platform Chemicals from Sustainable Resources*, Fang, Z., Smith, Jr., R. L., and Qi, X. (eds.), Springer, Singapore, pp. 413-436.

37. Abdelrahman, O. A., Luo, H. Y., Heyden, A., Roman-Leshkov, Y., and Bond, J. Q. (2015) Toward rational design of stable, supported metal catalysts for aqueous-phase processing: Insights from the hydrogenation of levulinic acid. *Journal of Catalysis*, **329**, 10-21.

38. Kruse, A., Funke, A., and Titirici, M. M. (2013) Hydrothermal conversion of biomass to fuels and energetic materials. *Current Opinion in Chemical Biology*, **17**, 515-521.
39. Yan, K., Wu, G., Lafleur, T., Jarvis, C. (2014) Production, properties and catalytic hydrogenation of furfural to fuel additives and value-added chemicals. *Renewable and Sustainable Energy Reviews*, **38**, 663-676.
40. Lange, J.-P., van der Heide, E., van Buijtenen, J., and Price, R. (2012) Furfural - A promising platform for lignocellulosic biofuels. *ChemSusChem*, **5**, 150-166.
41. Li, X., Jia, P., and Wang, T. (2016) Furfural: A promising platform compound for sustainable production of C4 and C5 chemicals. *ACS Catalysis*, **6**, 7621-7640.
42. Bozell, J. J., Moens, L., Elliott, D. C., Wang, Y., Neuenscwander, G. G., Fitzpatrick, S. W., Bilski, R. J., and Jarnefeld, J. L. (2000) Production of levulinic acid and use as a platform chemical for derived products. *Resources, Conservation and Recycling*, **28**, 227-239.
43. Chalid, M., Heeres, H. J., and Broekhuis, A. A. (2012) Green polymer precursors from biomass-based levulinic acid. *Procedia Chemistry*, **4**, 260-267.
44. Jackson, M. A., White, M. G., Haasch, R. T., Peterson, S. C., and Blackburn, J. A. (2018) Hydrogenation of furfural at the dynamic Cu surface of $CuOCeO_2/Al_2O_3$ in a vapor phase packed bed reactor. *Molecular Catalysis*, **445**, 124-132.
45. Yan, K., Lafleur, T., and Liao, J. (2013) Facile synthesis of palladium nanoparticles supported on multi-walled carbon nanotube for efficient hydrogenation of biomass-derived levulinic acid. *Journal of Nanoparticle Research*, **15**, 1-7.
46. Yan, K., Lafleur, T., Wu, G., Liao, J., Ceng, C., and Xie, X. (2013) Highly selective production of value-added γ-valerolactone from biomass-derived levulinic acid using the robust Pd nanoparticles. *Applied Catalysis A: General*, **468**, 52-58.
47. Delidovich, I., and Palkovits, R. (2016) Catalytic isomerization of biomass-derived aldoses: A review. *ChemSusChem*, **9**, 547-561.
48. Rinaldi, R., Palkovits, R., and Schuth, F. (2008) Depolymerization of cellulose using solid catalysts in ionic liquids. *Angewandte Chemie, International Edition*, **47**, 8047-8050.
49. Luo, H., Klein, I. M., Jiang, Y., Zhu, H., Liu, B., Kenttamaa, H. I., and Abu-Omar, M. M. (2016) Total utilization of miscanthus biomass, lignin and carbohydrates, using earth abundant nickel catalyst. *ACS Sustainable Chemistry & Engineering*, **4**, 2316-2322.
50. Pileidis, F. D., and Titirici, M.-M. (2016) Levulinic acid biorefineries: new challenges for efficient utilization of biomass. *ChemSusChem*, **9**, 562-582.
51. Morone, A., Apte, M., and Pandey, R. A. (2015) Levulinic acid produc-

tion from renewable waste resources: Bottlenecks, potential remedies, advancements and applications. *Renewable and Sustainable
Energy Reviews*, **51**, 548-565.

52. Xu, Y. Y., and Boeing, W. J. (2013) Mapping biofuel field: A biblio-
metric evaluation of research output. *Renewable and Sustainable
Energy Reviews*, **28**, 82-91.

53. Maity, S. K. (2015) Opportunities, recent trends and challenges of
integrated biorefinery: Part I. *Renewable and Sustainable Energy Reviews*, **43**, 1427-1445.

54. Yan, K., Lafleur, T., Jarvis, C., and Wu, G. (2014) Clean and selective
production of γ-valerolactone from biomass-derived levulinic acid
catalyzed by recyclable Pd nanoparticle catalyst. *Journal of Cleaner
Production*, **72**, 230-232.

55. Hoydonckx, H. E., Van Rhijn, W. M., Van Rhijn, W. , De Vos, D. E., and
Jacobs, P. A. (2007) Furfural and derivatives. In: Ullmann's Encyclo-
pedia of Industrial Chemistry (ed.), Wiley., Germany,
doi:10.1002/14356007.a12 119.pub2

56. Horvath, I. T., Mehdi, H., Fabos, V., Boda, L., and Mika, L. T. (2008)
[gamma]-Valerolactone-a sustainable liquid for energy and carbon-
based chemicals. *Green Chemistry*, **10**(2), 238-242.

57. Fabos, V., Koczo, G., Mehdi, H., Boda, L., and Horvath, I. T. (2009) Bio-
oxygenates and the peroxide number: a safety issue alert. *Energy &
Environmental Science*, **2**, 767-769.

58. Rao, R., Dandekar, A., Baker, R. T. K., and Vannice, M. A. (1997) Properties of copper chromite catalysts in hydrogenation reactions. *Journal of Catalysis*, **171**, 406-419.

59. Yan, K., and Chen, A. (2013) Efficient hydrogenation of biomass-derived furfural and levulinic acid on the facilely synthesized noble-
metal-free Cu–Cr catalyst. *Energy*, **58**, 357-363.

60. Yan, K., Jarvis, C., Lafleur, T., Qiao, Y., and Xie, X. (2013) Novel synthesis of Pd nanoparticles for hydrogenation of biomass-derived
platform chemicals showing enhanced catalytic performance. *RSC
Advances*, **3**, 25865-25871.

61. Yan, K., Liao, J., Wu, X., and Xie, X. (2013) A noble-metal free Cu-cat-
alyst derived from hydrotalcite for highly efficient hydrogenation of
biomass-derived furfural and levulinic acid. *RSC Advances*, **3**, 3853-
3856.

62. Lange, J.-P., and Van Buijtenen, J. (2011) Process for the Hydrogenolysis of Furfuryl Derivatives, US patent 20110184195A1.

63. Bremner, J. G., and Keeys, R. K. (1947) The hydrogenation of furfur-
aldehyde to furfuryl alcohol and sylvan (2-methylfuran). *Journal of
the Chemical Society*, 1068-1080.

64. Yan, K., Liu, Y., Lu, Y., Chai, J., and Sun, L. (2017) Catalytic application
of layered double hydroxide-derived catalysts for the conversion of
biomass-derived molecules. *Catalysis Science & Technology*, **8**,

1622-1645.

65. Hronec, M., and Fulajtarova, K. (2012) Selective transformation of furfural to cyclopentanone. *Catalysis Communications*, **24**, 100-104.

66. Yu, W., Tang, Y., Mo, L., Chen, P., Lou, H., and Zheng, X. (2011) One-step hydrogenation-esterification of furfural and acetic acid over bi-functional Pd catalysts for bio-oil upgrading. *Bioresource Technology*, **102**, 8241-8246.

67. Hronec, M., Fulajtarova, K., and Liptaj, T. (2012) Effect of catalyst and solvent on the furan ring rearrangement to cyclopentanone., *Applied Catalysis A: General*, **437-438**, 104-111.

68. Yang, J., Zheng, H.-Y., Zhu, Y.-L., Zhao, G.-W., Zhang, C.-H., Teng, B.-T., Xiang, H.-W., and Li, Y. (2004) Effects of calcination temperature on performance of Cu-Zn-Al catalyst for synthesizing [gamma]-butyr-olactone and 2-methylfuran through the coupling of dehydrogena-tion and hydrogenation. *Catalysis Communications*, **5**, 505-510.

69. Kijenski, J., Winiarek, P., Paryjczak, T., Lewicki, A., and Mikolajska, A. (2002) Platinum deposited on monolayer supports in selective hy-drogenation of furfural to furfuryl alcohol. *Applied Catalysis A: General*, **233**, 171-182.

70. Zheng, H.-Y., Zhu, Y.-L., Huang, L., Zeng, Z.-Y., Wan, H.-J., and Li, Y.-W. (2008) Study on Cu-Mn-Si catalysts for synthesis of cyclohexanone and 2-methylfuran through the coupling process. *Catalysis Communications*, **9**, 342-348.

71. Zheng, H.-Y., Zhu, Y.-L., Teng, B.-T., Bai, Z.-Q., Zhang, C.-H., Xiang, H.-W., and Li, Y.-W. (2006) Towards understanding the reaction path-way in vapour phase hydrogenation of furfural to 2-methylfuran. *Journal of Molecular Catalysis A: Chemical*, **246**, 18-23.

72. Yan, K., and Chen, A. (2014) Selective hydrogenation of furfural and levulinic acid to biofuels on the ecofriendly Cu-Fe catalyst. *Fuel*, **115**, 101-108.

73. Huang, W., Li, H., Zhu, B., Feng, Y., Wang, S., and Zhang, S. (2007) Se-lective hydrogenation of furfural to furfuryl alcohol over catalysts prepared via sonochemistry. *Ultrasonics Sonochemistry*, **14**, 67-74.

74. Zheng, H.-Y., Zhu, Y.-L., Bai, Z.-Q., Huang, L., Xiang, H.-W., and Li, Y.-W. (2006) An environmentally benign process for the efficient syn-thesis of cyclohexanone and 2-methylfuran. *Green Chemistry*, **8**, 107-109.

75. Yang, X. H., Xiang, X. M., Chen, H. M., Zheng, H. Y., Li, Y. W., and Zhu, Y. L. (2017) Efficient synthesis of furfuryl alcohol and 2-methylfu-ran from furfural over mineral-derived Cu/ZnO catalysts. *Chem-CatChem*, **9**, 3023-3030.

76. Gong, W., Chen, C., Zhang, H., Wang, G., and Zhao, H. (2017) Efficient synthesis of 2-methylfuran from bio-derived furfural over sup-ported copper catalyst: The synergistic effect of CuOx and Cu. *ChemistrySelect*, **2**, 9984-9991.

77. Srivastava, S., Jadeja, G. C., and Parikh, J. (2016) A versatile bi-metallic copper–cobalt catalyst for liquid phase hydrogenation of furfural to 2-methylfuran. *RSC Advances*, **6**, 1649-1658.
78. Fu, Z., Wang, Z., Lin, W., Song, W., and Li, S. (2017) High efficient conversion of furfural to 2-methylfuran over Ni-Cu/Al_2O_3 catalyst with formic acid as a hydrogen donor. *Applied Catalysis A: General*, **547**, 248-255.
79. Wang, B., Li, C., He, B., Qi, J., and Liang, C. (2017) Highly stable and selective Ru/NiFe 2 O 4 catalysts for transfer hydrogenation of biomass-derived furfural to 2-methylfuran. *Journal of Energy Chemistry*, **26**, 799-807.
80. Date, N. S., Biradar, N. S., Chikate, R. C., and Rode, C. V. (2017) Effect of reduction protocol of Pd catalysts on product distribution in furfural hydrogenation. *ChemistrySelect*, **2**, 24-32.
81. Chang, X., Liu, A. F., Cai, B., Luo, J. Y., Pan, H., and Huang, Y. B. (2016) Catalytic transfer hydrogenation of furfural to 2-methylfuran and 2-methyltetrahydrofuran over bimetallic copper-palladium catalysts. *ChemSusChem*, **9**, 3330-3337.
82. Stevens, J. G., Bourne, R. A., Twigg, M. V., and Poliakoff, M. (2010) Real-time product switching using a twin catalyst system for the hydrogenation of furfural in supercritical CO_2. *Angewandte Chemie, International Edition*, **122**(47), 9040-9043.
83. Yang, Y., Du, Z., Huang, Y., Lu, F., Wang, F., Gao, J., and Xu, J. (2013) Conversion of furfural into cyclopentanone over Ni-Cu bimetallic catalysts. *Green Chemistry*, **15**, 1932-1940.
84. Hronec, M., Fulajtarova, K., and Micucik, M. (2013) Influence of furanic polymers on selectivity of furfural rearrangement to cyclopentanone. *Applied Catalysis A: General*, **468**, 426-431.
85. Liu, S., Amada, Y., Tamura, M., Nakagawa, Y., and Tomishige, K. (2014) One-pot selective conversion of furfural into 1,5-pentanediol over a Pd-added Ir-ReOx/SiO_2 bifunctional catalyst. *Green Chemistry*, **16**, 617-626.
86. Yan, K., Wu, X., An, X., and Xie, X. (2013). Novel preparation of nanocomposite CuO-Cr_2O_3 using CTAB-template method and efficient for hydrogenation of biomass-derived furfural. *Functional Materials Letters*, **6**(1).
87. Gowda, A. S., Parkin, S., and Ladipo, F. T. (2012) Hydrogenation and hydrogenolysis of furfural and furfuryl alcohol catalyzed by ruthenium (II) bis (diimine) complexes. *Applied Organometallic Chemistry*, **26**, 86-93.
88. Biradar, N. S., Hengne, A. M., Birajdar, S. N., Niphadkar, P. S., Joshi, P. N., and Rode, C. V. (2014) Single-pot formation of THFAL via catalytic hydrogenation of FFR over Pd/MFI Catalyst. *ACS Sustainable Chemistry & Engineering*, **2**, 272-281.
89. Seo, G., and Chon, H. (1981) Hydrogenation of furfural over copper-

containing catalysts. *Journal of Catalysis*, **67**, 424-429.

90. Nakagawa, Y., Nakazawa, H., Watanabe, H., and Tomishige, K. (2012) Total hydrogenation of furfural over a silica-supported nickel catalyst prepared by the reduction of a nickel nitrate precursor. *Chem-CatChem*, **4**, 1791-1797.

91. Khairi, S., Hara, T., Ichikuni, N., and Shimazu, S. (2012) Highly efficient and selective hydrogenation of unsaturated carbonyl compounds using Ni–Sn alloy catalysts. *Catalysis Science & Technology*, **2**, 2139-2145.

92. Nakagawa, Y., and Tomishige, K. (2010) Total hydrogenation of furan derivatives over silica-supported Ni-Pd alloy catalyst. *Catalysis Communications*, **12**, 154-156.

93. Merat, N., Godawa, C., and Gaset, A. (1990) High selective production of tetrahydrofurfuryl alcohol: Catalytic hydrogenation of furfural and furfuryl alcohol. *Journal of Chemical Technology & Biotechnology*, **48**, 145-159.

94. Ordomsky, V.V., Schouten, J., Van Der Schaaf, J., and Nijhuis, T. (2013) Biphasic single-reactor process for dehydration of xylose and hydrogenation of produced furfural. *Applied Catalysis A: General*, **451**, 6-13.

95. Guo, H., Zhang, H., Zhang, L., Wang, C., Peng, F., Huang, Q., Xiong, L., Huang, C., Ouyang, X., Chen, X., and Qiu, X. (2018) Selective hydrogenation of furfural to furfuryl alcohol over acid-activated attapulgite-supported NiCoB amorphous alloy catalyst. *Industrial & Engineering Chemistry Research*, **57**, 498-511.

96. Rogowski, J., Andrzejczuk, M., Berlowska, J., Binczarski, M., Kregiel, D., Kubiak, A., Modelska, M., Szubiakiewicz, E., Stanishevsky, A., Tomaszewska, J., and Witonska, I. (2017) WxC-β-SiC nanocomposite catalysts used in aqueous phase hydrogenation of furfural. *Molecules*, **22**(11), 2033.

97. Yin, D., Ren, H., Li, C., Liu, J., and Liang, C. (2018) Highly selective hydrogenation of furfural to tetrahydrofurfuryl alcohol over MIL-101(Cr)-NH$_2$ supported Pd catalyst at low temperature. *Chinese Journal of Catalysis*, **39**, 319-326.

98. Li, C., Xu, G., Liu, X., Zhang, Y., and Fu, Y. (2017) Hydrogenation of biomass-derived furfural to tetrahydrofurfuryl alcohol over hydroxyapatite-supported Pd catalyst under mild conditions. *Industrial & Engineering Chemistry Research*, **56**, 8843-8849.

99. Messori, M., and Vaccari, A. (1994) Reaction pathway in vapor phase hydrogenation of maleic anhydride and its esters to γ-butyrolactone. *Journal of Catalysis*, **150**, 177-185.

100. Kanetaka, J., Asano, T., and Masamune, S. (1970) New process for production of tetrahydrofuran. *Industrial & Engineering Chemistry*, **62**, 24-32.

101. Zeitsch, K. J. (2000) *The Chemistry and Technology of Furfural and*

its *Many By-Products*, 13th volume, Elsevier, USA.

102. Yan, K., Liao, J., Wu, X., and Xie, X. (2013) Facile synthesis of eco-friendly Cu-hydrotalcite catalysts for highly selective synthesis of furfural diethyl acetal and benzoin ethyl ether. *Advanced Materials Letters*, **4**, 702-707.

103. Gilbert, W. W., and Howk, B. W. (1956) Process of Hydrogenating Maleic Anhydride with a Nickel or Cobalt Molybdite Catalyst, US patent US2772293A.

104. Rackemann, D. W., and Doherty, W. O. (2011) The conversion of lignocellulosics to levulinic acid. *Biofuels, Bioproducts and Biorefining*, **5**(2), 198-214.

105. Piskun, A. S., Ftouni, J., Tang, Z., Weckhuysen, B. M., Bruijnincx, P. C. A., and Heeres, H. J. (2018) Hydrogenation of levulinic acid to γ-valerolactone over anatase-supported Ru catalysts: Effect of catalyst synthesis protocols on activity. *Applied Catalysis A: General*, **549**, 197-206.

106. Horvath, A., Masanet, E., McKone, T., Lobscheid, A., Mishra, U., Fingerman, K., Lipman, T., and Auffhammer, M. (2010) Large-scale advanced biofuel implementation: A case study of Illinois and Indiana. *ACS National Meeting Book of Abstracts 239*, USA.

107. Fabos, V., Mika, L. T., and Horvath, I. T. (2014) Selective conversion of levulinic and formic acids to γ-valerolactone with the Shvo catalyst. *Organometallics*, **33**, 181-187.

108. Fegyverneki, D., Orha, L., Láng, G., and Horvath, I. T. (2010) Gamma-valerolactone-based solvents. *Tetrahedron*, **66**, 1078-1081.

109. Geilen, F. M. A., Engendahl, B., Holscher, M., Klankermayer, J., and Leitner, W. (2011) Selective homogeneous hydrogenation of biogenic carboxylic acids with [Ru(TriPhos)H]+: A mechanistic study. *Journal of the American Chemical Society*, **133**, 14349-14358.

110. Yan, K., Lafleur, T., Liao, J., and Xie, X. (2014) Facile green synthesis of palladium nanoparticles for efficient liquid-phase hydrogenation of biomass-derived furfural. *Science of Advanced Materials*, **6**, 135-140.

111. Wang, A. Lu, Y., Yi, Z., Ejaz, A., Hu, K., Zhang, L., and Yan, K. (2018) Selective production of γ-valerolactone and valeric acid in one-pot bifunctional metal catalysts. *ChemistrySelect*, **3**, 1097-1101.

112. Manzer, L. E. (2004) Catalytic synthesis of [alpha]-methylene-[gamma]-valerolactone: a biomass-derived acrylic monomer. *Applied Catalysis A: General*, **272**, 249-256.

113. Bourne, R. A., Stevens, J. G., Ke, J., and Poliakoff, M. (2007) Maximising opportunities in supercritical chemistry: the continuous conversion of levulinic acid to c-valerolactone in CO_2. *Chemical Communications*, 4632-4634.

114. Deng, L., Zhao, Y., Li, J., Fu, Y., Liao, B., and Guo, Q.-X. (2010) Conversion of levulinic acid and formic acid into γ-valerolactone over het-

erogeneous catalysts. *ChemSusChem*, **3**, 1172-1175.

115. Yan, Z.-p., Lin, L., and Liu, S. (2009) Synthesis of γ-valerolactone by hydrogenation of biomass-derived levulinic acid over Ru/C catalyst. *Energy & Fuels*, **23**, 3853-3858.

116. Kopetzki, D., and Antonietti, M. (2010) Transfer hydrogenation of levulinic acid under hydrothermal conditions catalyzed by sulfate as a temperature-switchable base. *Green Chemistry*, **12**, 656-660.

117. Upare, P. P., Lee, J. -M., Hwang, D. W., Halligudi, S. B., Hwang, Y. K., and Chang, J.-S. (2011) Selective hydrogenation of levulinic acid to [gamma]-valerolactone over carbon-supported noble metal catalysts. *Journal of Industrial and Engineering Chemistry*, **17**, 287-292.

118. Du, X. -L., Bi, Q. -Y., Liu, Y. -M., Cao, Y., and Fan, K. -N. (2011) Conversion of biomass-derived levulinate and formate esters into γ-valerolactone over supported gold catalysts. *ChemSusChem*, **4**, 1838-1843.

119. Du, X. -L., He, L., Zhao, S., Liu, Y.-M., Cao, Y., He, H.-Y., and Fan, K. -N. (2011) Hydrogen-independent reductive transformation of carbohydrate biomass into γ-valerolactone and pyrrolidone derivatives with supported gold catalysts. *Angewandte Chemie, International Edition*, **50**, 7815-7819.

120. Son, P. A., Nishimura, S., and Ebitani, K. (2014) Production of [gamma]-valerolactone from biomass-derived compounds using formic acid as a hydrogen source over supported metal catalysts in water solvent. *RSC Advances*, **4**, 10525-10530.

121. Yang, Y., Gao, G., Zhang, X., and Li, F. (2014) Facile fabrication of composition-tuned Ru–Ni bimetallics in ordered mesoporous carbon for levulinic acid hydrogenation. *ACS Catalysis*, **4**, 1419-1425.

122. Sudhakar, M., Lakshmi Kantam, M., Swarna Jaya, V., Kishore, R., Ramanujachary, K. V., and Venugopal, A. (2014) Hydroxyapatite as a novel support for Ru in the hydrogenation of levulinic acid to γ-valerolactone. *Catalysis Communications*, **50**, 101-104.

123. Al-Shaal, M. G., Dzierbinski, A., and Palkovits, R. (2014) Solvent-free [gamma]-valerolactone hydrogenation to 2-methyltetrahydrofuran catalysed by Ru/C: A reaction network analysis. *Green Chemistry*, **16**, 1358-1364.

124. Geilen, F., Engendahl, B., Harwardt, A., Marquardt, W., Klankermayer, J., and Leitner, W. (2010) Selective and flexible transformation of biomass-derived platform chemicals by a multifunctional catalytic system. *Angewandte Chemie, International Edition*, **49**, 5510-5514.

125. Hayes, D. J. (2009) An examination of biorefining processes, catalysts and challenges. *Catalysis Today*, **145**, 138-151.

126. Nandiwale, K. Y., and Bokade, V. V. (2014) Environmentally benign catalytic process for esterification of renewable levulinic acid to various alkyl levulinates biodiesel. *Environmental Progress & Sustai-*

nable Energy, **34**(3), 795-801.

127. Yan, K., Wu, G., Wen J., and Chen, A. (2013) One-step synthesis of mesoporous $H_4SiW_{12}O_{40}$-SiO_2 catalysts for the production of methyl and ethyl levulinate biodiesel. *Catalysis Communications*, **34**, 58-63.

128. Nandiwale, K. Y., Pande, A. M., and Bokade, V. V. (2015) One step synthesis of ethyl levulinate biofuel by ethanolysis of renewable furfuryl alcohol over hierarchical zeolite catalyst. *RSC Advances*, **5**, 79224-79231.

129. Nandiwale, K. Y., Sonar, S. K., Niphadkar, P. S., Joshi, P. N., Deshpande, S. S., Patil, V. S., and Bokade, V. V. (2013) Catalytic upgrading of renewable levulinic acid to ethyl levulinate biodiesel using dodecatungstophosphoric acid supported on desilicated H-ZSM-5 as catalyst. *Applied Catalysis A: General*, **460-461**, 90-98.

130. Pasquale, G., Vazquez, P., Romanelli, G., and Baronetti, G. (2012) Catalytic upgrading of levulinic acid to ethyl levulinate using reusable silica-included Wells-Dawson heteropolyacid as catalyst. *Catalysis Communications*, **18**, 115-120.

131. Patil, M. K., Keller, M., Reddy, B. M., Pale, P., and Sommer, J. (2008) Copper zeolites as green catalysts for multicomponent reactions of Aldehydes, terminal alkynes and amines: An efficient and green synthesis of propargylamines. *European Journal of Organic Chemistry*, **2008**, 4440-4445.

132. Fernandes, D. R., Rocha, A. S., Mai, E. F., Mota, C. J. A., and Teixeira da Silva, V. (2012) Levulinic acid esterification with ethanol to ethyl levulinate production over solid acid catalysts. *Applied Catalysis A: General*, **425-426**, 199-204.

133. Oliveira, B. L., and Teixeira da Silva, V. (2014) Sulfonated carbon nanotubes as catalysts for the conversion of levulinic acid into ethyl levulinate. *Catalysis Today*, **234**, 257-263.

134. Barau, A., Budarin, V., Caragheorgheopol, A., Luque, R. Macquarrie, D., Prelle A., Teodorescu, V., and Zaharescu, M. (2008) A Simple and efficient route to active and dispersed silica supported palladium nanoparticles. *Catalysis Letters*, **124**, 204-214.

135. Li, H., Fang, Z., Luo, J., and Yang, S. (2017) Direct conversion of biomass components to the biofuel methyl levulinate catalyzed by acid-base bifunctional zirconia-zeolites. *Applied Catalysis B: Environmental*, **200**, 182-191.

Chapter 3

Biosynthetic Pathway for Production of Renewable Biofuels

Happy Panchasara,[§] Shreya Patel,[§] Nisarg Gohil and Vijai Singh*
Synthetic Biology Laboratory, Department of Microbiology, School of Biological Sciences and Biotechnology, Institute of Advanced Research, Koba Institutional Area, Gandhinagar 382426, India
Corresponding author: vijai.singh@iar.ac.in
[§]Both authors contributed equally to this work

3.1 Introduction

The global demand of fuels has significantly increased, while fossils fuels are depleting. The industrial revolution has inescapably relied on fossil fuels and our consumption of fossils has exponentially increased. Amongst all other energy sources, oil is predominantly consumed energy source. Entailing the crude oil and its products such as gasoline, diesel, and kerosene as major energy source has forged issues of environmental concern such as pollution and global warming. Besides this, circumstances of energy and economic crisis also originate due to fossil fuels. Major reasons leading to such issues of global stress include CO_2 emission, fossil fuel depletion and peaked fuel prices. The CO_2 emission is caused by the combustion of fuels that has adverse effect on the environment and animal health. As per International Energy Agency, the total greenhouse gas emissions is the highest from the oxidation of carbon present in the fuels. The concentration of CO_2 in the atmosphere prior to the industrial era was about 280 parts per million (ppm) which has increased to 403 ppm in 2016. The average increase in the concentration of CO_2 in the last ten years was 2 ppm per year [1]. Anthropogenic emission of CO_2 and other greenhouse gases consequently fosters detrimental environmental and health impacts. Thus, it is high time to develop and switch to low carbon fuels in order to protect the environment. Another important factor that restricts the continuous consumption of fossil fuels is that these are obtained from earth's crust and heavy utilization of these resources is resulting in their exhaustion. The

Biofuels, edited by Vikas Mittal
© 2018 Central West Publishing, Australia

rate of extraction of these resources has already slowed down meaning that they are significantly depleting. Due to the decline in the supply of fossils with respect to the demand, an imbalance in demand and supply is generated which affects the fuel prices. Apart from demand and supply, another significant factor that fluctuates the prices of fossil fuels is the prevalent socio-political situation worldwide. As fossil fuels are a non-renewable form of energy that is not replenished, relentless utilization would result in the fossil impoverishment. If such unsustainable use of fossils continues, it would leave no resources for future generations. Several reports have forecasted the future production of fossil fuels. For instance, Mohr *et al.* [2] evaluated Ultimately Recoverable Resources (URR), based on the total amount of recoverable resources from the available ground resources. According to their estimation, the fossil fuel production will rapidly decline after 2025. Another study determined the number of years oil, coal and gas reserves will last on the basis of ratio of world consumption and reserves. According to this study, oil, coal, and gas would deplete in 35, 107 and 37 years respectively, indicating that oil will be depleted sooner than the other two energy types [3].

Although the precise prediction of fossil fuel reserves and their depletion time is difficult as the consumption depends on numerous factors, however, it can be conjectured that these will be depleted swiftly as our dependence on fuels for economic growth is incessant. Therefore, in order to safeguard the fossil fuel reserves, it is important to substitute non-renewable form of energy by renewable forms of energy. Thus, a pressing need has arisen to identify the alternatives to fossil fuels. Biofuels are a safer, cleaner and ecological substitute, through the price of biofuels is currently high. However, the future technological developments will allow us to produce these in a renewable and sustainable manner with the aid of metabolic engineering and synthetic biology.

A number of microorganisms have natural ability to produce biofuels that can be used to replace the fossil fuels. The properties of major biofuels such as ethanol, biodiesel and butanol make them more suitable than other sort of fuels. Considering the fact that these are produced from biomass which can reduce the atmospheric CO_2 by the process of photosynthesis, thus, these biofuels are environmental friendly. Besides this, the oxygen content of biofuels allows for their complete combustion which consequently lessens the CO_2 emission, unburned hydrocarbons and particulate matter emission, thus, imparting safety to the environment and human health [4,5]. Another related advantage is the higher octane number of alcohol derived fuels in comparison to conventional fuels, which resists the knocking of engines. Biomethanol, bioethanol and biobutanol are some of the alcohol derived biofuels used as substitutes for fossil fuels and have high octane number. Though ethanol and methanol can be used as biofuels, however, butanol is more suitable due to its similarities with gasoline. Butanol has also longer hydrocarbon chain than ethanol and methanol. It has higher energy content which makes it ideal for use in combustion engines without any modifications. Unlike ethanol, which absorbs water from atmosphere, thus, making it corrosive in nature, butanol has lower water absorption that allows better combustion engine life [6].

Butanol seems to be the potential future fuel that can be gradually used to replace the fossil fuels. Principally, there are two predominant ways to produce butanol: first is from the biomass that involves the conversion of cellulosic feedstock into fuel by microorganisms and second from fossil fuel products like petrol. Butanol produced from petroleum is called petro-butanol and its value is nullified because of its raw material. The most appropriate method for large scale generation of butanol for commercial purposes is by microbial production. Some microbes naturally produce butanol by fermentation of sugars. However, the natural or wild-type strains encounter various challenges including substrate specificity, by-product formation (such as ethanol and acetone production in certain butanol producing strains), low titre and product toxicity [7]. These natural strains lack certain genetic modifications and a suitable cell physiology is necessary for high product titre. Also, it is hard to manipulate their genetic system due to lack of molecular biology tools. Therefore, heterologous hosts are chosen as an alternative way to produce larger amounts of butanol by extension or modification of biosynthetic pathway. Microbial strain development can eliminate the constraints a natural strain confronts and a strain with higher substrate affinity, lower by-product formation, optimized metabolic fluxes, increased productivity and improved tolerance for product or by-product can be obtained [8]. To produce robust strains, different strategies can be employed such as mutagenesis, omics, directed evolution and synthetic biology [7,9].

Synthetic biology tools facilitate the engineering of microorganisms for the production of commercially important products. Application of synthetic biology tools assists in metabolic engineering of microorganisms by optimizing metabolic flux via modification of existing pathways, designing novel pathways for biofuel production or transferring existing pathway in different organism and, thereby, producing high energy, economical and eco-friendly fuels [10-13]. In the present chapter, we highlight the recent advances and future opportunities for producing low-cost renewable biofuels at industrial scale by expression of biosynthetic pathways in bacteria, yeast, algae and cyanobacteria.

3.2 Natural Microbial Production of Butanol

Clostridium is one of the well-known microorganisms which has natural ability to produce butanol. Its ability to produce solvents by fermentation of sugars has been scrutinized to a large extent by researchers. Though, clostridia species can produce butanol, nevertheless, their ability needs to be enhanced by the manipulation of biosynthetic pathways in order to improve productivity. The flow of carbon is the principal feature which decides the fate of a product. Therefore, manipulating the genes for redirecting the carbon flux is a way to improve the production [14]. The fermentation of sugars in clostridia occurs in two phases: acidogenesis and solventogenesis. Sugars are initially fermented to form acids such as acetic acid and butyryl which lowers the pH of medium. Increased concentration of acidic products is toxic to the cells, and cells consequently start converting the previously accumulated acids to solvents in particular forms such as butanol, acetone and ethanol. Butyrate and butanol are formed from butyryl CoA, of which butyrate is the product of acid fermentation while butanol is the product of butanol fermentation. Conversion of butyryl CoA to butyrate requires two genes: *ptb* (converts butyryl CoA to butyryl phosphate) and *buk* (converts butyryl phosphate to butyrate). Knocking out genes, which encode enzymes, responsible for acid formation is a strategy by which acidogenesis can be prevented and solventogenesis can be elevated. For instance, Harris *et al.* [15] reported the altered production of alcohols by the inactivation of *buk* (butyrate kinase). The gene inactivation decreased the production of butyrate, accordingly accumulating butyryl CoA. Due to butanol formation, the flux increased by up to 300 % as compared to wild-type strain. A list of butanol producing wild-type and engineered strains is presented in Table 3.1. The table also contains the involved pathways, genes, mechanism and titre of butanol.

Table 3.1 Summary of the engineered microorganisms for butanol production

Organism	Pathway	Gene(s)	Mechanism	Titre	Ref
Clostridium acetobutylicum	*Clostridium* CoA pathway	*buk*	Selective gene inactivation using non-replicative plasmid	16.7 g/L	[15]
Clostridium acetobutylicum	*Clostridium* CoA pathway	*pta, buk*	Gene knockout	18.9 g/L	[16]
		adhE1^{D485G}	Overexpression		
Clostridium acetobutylicum	*Clostridium* CoA pathway	*solR*	Inactivation by homologus recombination	17.8 g/L	[17]
Clostridium acetobutylicum	*Clostridium* CoA pathway	*adc*	TargeTron gene knockout	13.6 g/L	[19]
Clostridium acetobutylicum	Glutathione synthesis	*gshAB*	Cloned and expressed using plasmid	14.8 g/L	[23]
Clostridium pasteurianum	Not applicable	*rex*	In-frame deletion	9.9 g/L	[22]
		hydA		7.79 g/L	
Clostridium tyrobutyricum	*Clostridium* CoA pathway	*ack*	Deletion	10 g/L	[21]
		adhE2	Overexpression		
Escherichia coli	*Clostridium* CoA pathway	*thl, hbd, crt, bcd, etfAB, adhE2*	Cloned and expressed using plasmids	552 mg/L	[25]
		pta, fnr	Deletion		

Escherichia coli	Clostridium CoA pathway	a to B, hbd, crt, adhE2, ter, fdh	Insertion	30 g/L	[27]
		pta	Deletion		
Escherichia coli	Clostridium CoA and Valine pathway	ilvIHCD, alsS	Overexpression	22 g/L	[28]
		pflB	Deletion		
Escherichia coli	Clostridium CoA pathway	adhE, frdBC, fnr, ldhA, pta, pflB	Deletion	50 g/L	[24]
Bacillus subtilis	Clostridium CoA and valine pathway	alsS, kvid, adh	Overexpression	2.62 g/L	[29]
Corynebacterium glutamicum	Valine pathway	ilvBNCD, kivd, adhA, pntAB	Overexpressed	12.9 g/L	[30]
Saccharomyces cerevisiae	Clostridium CoA pathway	thl, hbd, crt, bcd, etfAB, adhE2	Overexpression	2.5 mg/L	[34]
Saccharomyces cerevisiae	Valine pathway	kivd, adh	Overexpression	143 mg/L	[35]
		pdc1	Deletion		
Saccharomyces cerevisiae	Threonine pathway	leu1, leu4, leu2, leu5, cimA, nfs1, adh7, aro10	Overexpression	835 mg/L	[36]
Synechocystis elongatus PCC 7942	Clostridium CoA pathway	adh	Insertion	450 mg/L	[39]

Synechocystis PCC 6803	*Clostridium* CoA and Valine pathway	*kivd, ADH*	Insertion	60.8 mg/L	[40]
Synechocystis elongatus PCC 7942	*Clostridium* CoA pathway	*hbd, crt, adhE2, ter, atoB*	Insertion	14.5 mg/L	[42]
Synechococcus elongatus PCC7942	Valine pathway	*kivd, yqhD*	Insertion	200 mg/L	[47]
Synechocystis elongatus PCC 6803	Valine pathway	*kivd, adhA*	Insertion	90 mg/L (from bicarbonate, autotrophically) 114 mg/L (from glucose, autotrophically) 298 mg/L (mixotrophically)	[38]

Similarly, activity of two genes *pta* and *buk* encoding the enzymes phosphotransacetylase and butyrate kinase was disrupted and another gene *adhE1D485G* encoding a mutated aldehyde/alcohol dehydrogenase was overexpressed simultaneously [16]. The engineered strains produced 18.9 g/L of butanol which is higher than the wild-type strain. As mentioned earlier, butanol was formed by two ways: by butyrate accumulation, which is mentioned as a cold channel and by direct formation of butanol from butyryl CoA, which is known as a hot channel. Enzymes of both channels were observed to be individually knocked out as well as also in combination. The strain was characterized by metabolic flux and mass balance analysis. Based on these experiments, the authors confirmed that direct pathway (hot channel) significantly increased the yield of butanol [16]. Overall, overexpression of solventogenic genes or knockout of acid formation genes are important for bioengineering of clostridia to produce butanol.

In another study, Nair *et al.* [17] have demonstrated that early induction of the same genes is equally important for elevating the butanol titre. *SolR* is a putative transcriptional repressor that is associated with solvent production. During the acidogenesis phase, *solR* was expressed, which repressed the solventogenic genes at the early exponential phase. On the other hand, in the late exponential phase, the changes in the medium lead to the induction of solventogenic genes. The authors demonstrated the increased butanol yield due to inactivation of *solR* gene as rendering it inactive allows for early induction of solventogenic genes.

The strategy of redirecting the carbon flux by inactivation or down-regulation of certain acid formation genes manifests enhanced solvent production. Nevertheless, the solventogenesis process produces alcohols (butanol, acetone and ethanol) in a definite ratio of 6:3:1 respectively, in different species. Therefore, improving the selectivity of butanol production by manipulating the biosynthetic pathway is another important objective of strain improvement to enhance the production of butanol and simplify the final product recovery. Since acetone is produced from acetoacetate by enzyme acetoacetate decarboxylase (AADC) which is encoded by *adc* and acetoacetate is formed from acetyl-CoA by coenzyme A-transferase (CoAT) composed of two units A and B that are encoded by *ctfA* and *ctfB*, thus, the modification in the gene may alter the acetone production. In a study, Tummala *et al.* [18] used antisense RNA technique to initially down-regulate the AADC. The down-regulation of AADC did not collaterally affect the acetone production. On the basis of this finding, the authors reported that the AADC enzyme was not the rate-limiting factor in acetone production. It means that AADC is important for producing acetone, but even low levels of the enzyme are enough for production. Subsequently, the authors targeted *CoAT* and observed that down-regulation of the enzyme by antisense RNA considerably decreased acetone production. The study determined the rate-limiting step of acetone forming reaction which is an important finding for improving butanol selectivity. It is important to exert influence on the rate-limiting step of acetone production as it is the largest produced byproduct after butanol. The antisense RNA strategy did not prove to be efficient for reducing the acetone production by down-regulation of *adc*. However, to overcome this limitation, Jiang *et al.* [19] disrupted the *adc* gene by using TargeTron approach to eliminate the acetone production. TargeTron technology utilizes mobile group II introns to generate a mutated copy of the target gene that can be transferred in the cell using a vector to knockout a gene. It was employed to inactivate *adc* gene so that acid accumulation takes place. The accumulated acid inhibits the conversion of acetyl-CoA to acid, consequently alleviating acetone production and increasing butanol production. The fermentative study of mutated strain was carried out in P2 medium. In the medium, the butanol ratio was increased from 70% to 80.05% and acetone concentration was reduced up to 0.21 g/L. In order to control pH and regulate electron flow, 1% calcium carbonate and methyl viologen were added into medium. In this medium, the butanol ratio was increased to 82±2%, while control strain was 74.3±0.4%.

Similarly, *Clostridium acetobutylicum* M5 strain is non-sporulating and non-solventogenic strain, which was complemented with megaplasmid pSOL1 containing alcohol/aldehyde dehydrogenase (*aad*) gene and acetone formation gene [20]. Since the strain has lost gene for solvent production, it is difficult to achieve high titre. In this study, the wild-type level of titre was achieved in the strain by plasmid which overexpressed *aad* due to presence of altered promoter *ptb*. The strains were also incapable of re-assimilating acids, thus, thiolase (*thl*) overexpression was combined with *aad* overexpression. The strain was able to reduce the acetate production, whereas increasing butyrate and butanol production, however, the butanol production was also lowered as compared to butyrate [20]. Thus, by regulating the genes involved in the production of by-products other than butanol during solventogenesis, butanol selectivity was improved and acetone production was lessened or completely eliminated.

For butanol production, *C. acetobutylicum* is a good source, along with many other species in this genus able to produce alcohols. Modification of these strains for improving the productivity of butanol has also been reported. Mutants of *Clostridium tyrobutyricum* were generated in a study by incorporating *adhE2* (aldehyde/alcohol dehydrogenase) gene [21]. Previously created *ptb* and *ack* (genes for butyrate and acetate formation respectively) knockout mutants of *C. tyrobutyricum* with the increased flux towards the butyrate formation were further investigated in this study by overexpressing butanol forming gene *adhE2*, which accounts for conversion of butyryl CoA to butanol. Further, butanol production in different strains was compared. The authors observed that the butanol titre was increased up to 10 g/L in *ack* mutants with overexpressed *adhE2*. In addition, the authors also studied the effect of reduced substrate on butanol production wherein even higher butanol titre (16 g/L) was obtained. This strain also furnished a supplementary advantage of no acetone forming genes, thus, butanol yield could be high [21].

Inactivating genes which maintain the electron flow and redox potential in the organism and, thereby, increase butanol production and reduce acetone generation were not successfully achieved in a model organism. Schwarz *et al.* [22] demonstrated increased butanol formation by generating in-frame deletion mutants of *Clostridium pasteurianum*. Two genes such as *rex* and *hydA* encoding redoxresponsive regulator and hydrogenase were deleted which resulted in increased butanol titre from 93.2 mM in wild-type to 133.3 mM and 105.1 mM in *rex* and *hydA* mutants. Both genes were involved in the solvent production. The authors also demonstrated the utilization of glycerol as a feedstock, which is a by-product of biodiesel production. The gene *DhaBCE*- glycerol dehydratase, associated with the metabolism of glycerol, was removed, however, this caused the low growth of strain in the poor medium.

In addition to metabolic flux and butanol selectivity, another factor which significantly influences the butanol titre is butanol toxicity. The genetic manipulation of microorganism is a way that allows to generate butanol tolerant strain. To overcome the solvent toxicity, *gshAB* gene producing glutathione from *Escherichia coli* was over-expressed in *C. acetobutylicum*. The mutant strain was found to be aero-tolerant and butanol tolerant. Glutathione is known to protect several enzymes *in E. coli*, hence, it might give similar protective effect in *Clostridium* by generating a strain with the ability to grow in micro-toxic conditions. The concentration of butanol by a strain with *gshAB* gene was 14.8 g/L, while the wild-type strain was 10.8 g/L. It suggested that glutathione imparted capability to resist solvent, resulting in higher titre value [23]. A pressing need arises to identify a novel strain, along with rigorous fine-tuning and gene regulation for enhancement of butanol production. Butanol toxicity can, thus, be improved by strategies like directed evolution or protein export pump engineering or global transcription factor engineering that allow to obtain mutants with novel phenotype and high tolerance along with faster growth rate.

3.3 Production of Butanol in Heterologous Hosts

Clostridia are native producers of solvents with a variety of practical drawbacks including slower growth rate, being obligate anaerobes and sporulating. On the other hand, heterologous expression of solventogenic genes is an alternative way that stipulates advantage on selectivity of butanol production. Engineering of microorganisms by inserting genes for butanol production that are not inherently present within heterologous host can facilitate to overcome the obstruction of acetone and ethanol as byproduct. Moreover, the generation of genetic manipulations in clostridia has constraints due to its complex metabolic flux and regulation, hence, developing a non-solventogenic strain to solventogenic is less cumbersome.

Researchers have engineered alternative microbial hosts to ease the production of desired biofuels. *E. coli* is one of well-known microorganisms with enriched tacking genetic tools, known metabolism and physiology. Thus, it has been used as a potential source for butanol production by constructing a novel butanol pathway. In a study, it was demonstrated that engineered *E. coli* produced butanol equivalent to 50 g/L [24]. The yield of butanol was observed to be substantially higher than the previously reported native organisms. In a similar study, *E. coli* was engineered to produce butanol [25]. Genes (*thl, hbd, crt, bcd, etfAB, adhE2*) for butanol production pathway from *C. acetobutylicum* were cloned and expressed in *E. coli*. The yield of butanol titre was observed to be 13.9 mg/L. In order to improve the production of butanol, overexpression of *C. acetobutylicum* thiolase gene and acetyl-CoA acetyltransferase from *E. coli* was performed, which could increase the yield up to 3-folds. In order to further improve butanol production, the competing pathways genes (*ldhA, adhE*, and *frdBC*) of acetyl-CoA and NADH pathway were deleted. The authors also replaced *adhE* gene of *E. coli* with *adhE2* of *C. acetobutylicum*, as the later one had much lower activity for acetyl-CoA. Deletion of *pta* removed acetate production, while *fnr* removal caused an increase in activity of pyruvate dehydrogenase which is important for the formation of acetyl-CoA from pyruvate. The butanol production was increased up to 373 mg/L, which was further increased up to 552 mg/L when the cells were grown in nutrient rich medium [25]. In addition, Nielsen *et al.* [26] expressed same genes, but individually, along with overexpression of formate dehydrogenase encoding gene of *Saccharomyces cerevisiae*. The glycolytic flux was increased by over-expressing glyceraldehyde-3 phosphate dehydrogenase of *E. coli* and butanol titres (580 mg/L) were increased as compared to previous studies. Besides, the prospect of *Pseudomonas putida* and *Bacillus subtilis* as heterologous hosts for butanol production was also studied. Butanol biosynthetic genes were expressed as a polycistronic gene in both *P. putida* and *B. subtilis*, for which the butanol titres were observed to be 120 mg/L and 24 mg/L, respectively.

In another study, Shen *et al.* [27] demonstrated how the driving forces of butanol production pathways can direct the flux towards butanol formation and increase its titre. The authors modified the clostridial butanol formation pathway for this purpose. The butanol formation pathway utilizes NADH and ferredoxin (Bcd-EtfAB complex) as a source of reducing power. Here, the authors modified the pathway in such a manner that only NADH could be used as reducing power. As accumulation of NADH is necessary to fulfil the requirement of butanol producing pathway, mixed acid fermentation reactions which consume NADH to produce other substances are needed to be deleted. To use NADH as the only driving force, the authors employed the strategy of linking an enzyme (*Ter*). The enzyme trans-enoyl-coenzyme A (*CoA*) reductase (*Ter*), highly specific for NADH, was used instead of butyryl-CoA dehydrogenase complex which used Bcd-EtfAB complex in catalysis of crotonyl-CoA reduction step. In glycolysis reaction, only two molecules of NADH are formed, while one molecule of butanol formation requires four NADH. Hence to balance the amount of NADH produced by glycolysis, formate dehydrogenase was over-expressed which yields two molecules of NADH for one molecule of formate. The flux towards acetate production was eliminated by deleting *pta* (phosphate acetyltransferase) (Figure 3.1). The high butanol titre obtained was equal to 30 g/L [27]. In the previous studies where driving forces such as NADH and acetyl CoA were not coupled with *Ter* and (Bcd-EtfAB complex), only a small amount of 373 mg/L of titre was obtained [25].

Production of alcohols from non-fermentative pathway in *E. coli* has also been reported by Atsumi *et al.* [28]. The authors proposed that expression of non-native pathways in a host may lead to metabolic imbalance and hence taking the advantage of innate pathways in a host is more relevant. On the basis of this strategy, the authors took advantage of amino acid pathways in which 2-keto acids are intermediates and undergo two-step reactions to produce alcohol. In this study, 2-keto acids were derived from glucose by overexpression of specific genes that re-routed pyruvate to amino acid biosynthesis pathway (for 2-keto acid intermediates production) and to further generate alcohols in *E. coli*. Two enzymes are needed for the

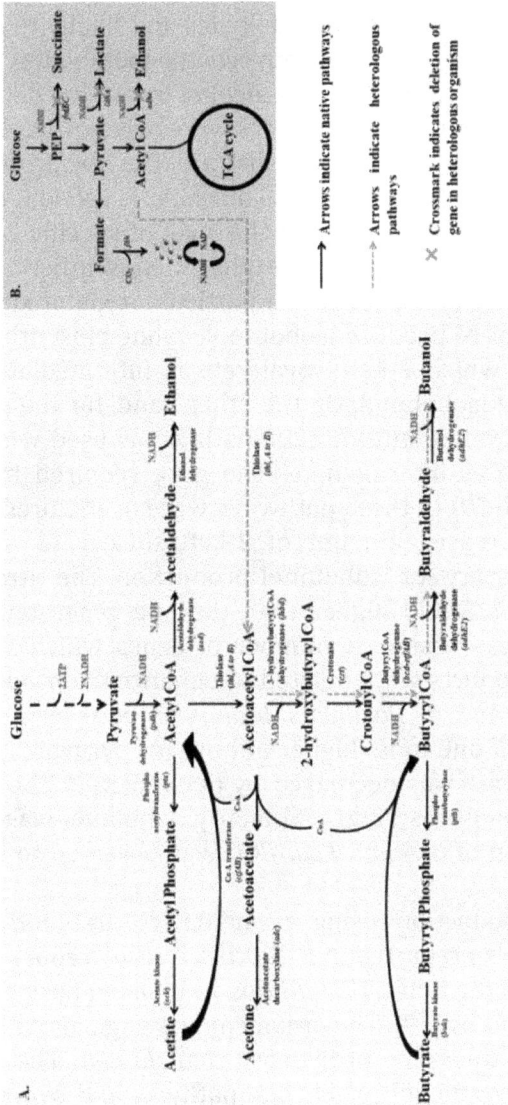

Figure 3.1 Schematic representation of a butanol production pathway (A) in *Clostridium*. (B) Metabolism in *E. coli*. The engineered pathway for butanol formation in *E. coli* involves six enzymatic reactions. The enzymes (*thl, hbd, crt,bcd-etfAB, adhE2*) for these reactions are cloned from *C. acetobutylicum* and expressed in *E. coli*. Native enzymes of *E. coli* involved in by-products formation are deleted. *frdBC, ldhA* and *adhE* encoding fumarate reductase, lactate dehydrogenase and alcohol dehydrogenase respectively are deleted to accumulate acetyl CoA and increase flux towards butanol formation. *fdh* encodes formate dehydrogenase.

conversion of 2-keto acids to aldehydes and finally to alcohols. The 2-keto decarboxylases (KDC) converts the metabolites to aldehydes which are further converted into alcohol by alcohol dehydrogenases (ADH). Five different types of KDC's were overexpressed in *E. coli* to verify their specificity with the substrate. The *Kivd* from *Lactococcus lactis* was capable of producing all alcohols. *E. coli* was subsequently genetically modified to increase the level of specific 2-keto acid so that the desired alcohol can be produced. Biosynthetic pathway of a particular amino acid produces a particular type of alcohol. For example, in order to produce isobutanol, valine biosynthesis pathway is utilized in which 2-ketoisovalerate is intermediate which gets converted into isobutanol. On the other hand, for the production of n-butanol, norvaline biosynthetic pathway is used which produces 2-ketovalerate as intermediate. The gene required to encode the enzymes (*ilvIHCD*) of these pathways were overexpressed in *E. coli*. Therefore, increased amount of 2-ketoisovalerate was produced which is necessary for isobutanol production. The titre of obtained isobutanol was 5-fold higher than the one generated from strain without overexpression. A number of genes which are known to produce byproducts were deleted to enhance the production of isobutanol. The *AlsS* of *Bacillus subtilis* was used as a substitute of *ilvIHC* of *E. coli* due to its higher affinity for pyruvate. The competition for pyruvate was decreased by deleting *pflB*. The yield of produced isobutanol was 22 g/L. Similarly, 1-butanol was produced by overexpression of *ilvA* and *LeuABCD* was observed to enhance titre [28].

Solvent production using fermentative and non-fermentative pathway are also reported in *B. subtilis*. A study reported 2.62 g/L of isobutanol production by *B. subtilis* in a shake-flask fed-batch fermentation process [29]. The organism was engineered because of its high product tolerance propensity than *E. coli*. The product was formed by introduction of Ehrlich pathway and overexpression of acetolactate synthase (*alsS*) which formed 2-ketoisovalerate and acetolactate, which are two important precursors for butanol formation. The overexpression of *alsS* enhanced acetolactate production, which is an important precursor for the formation of 2-ketoisovalerate. Further, by introducing genes of Ehrlich pathway, 2-ketoisovalerate was converted to isobutanol [29].

Other heterologous organisms in which butanol biosynthesis was reported by manipulation of native pathways include *Corynebacterium glutamicum*. A mutant strain of *C. glutamicum* was generated and characterized for butanol formation [30]. Several genes were deleted including *aceE, pqo, ilvE, ldhA* and *mdh* that encode E1p subunit of the pyruvate dehydrogenase complex, pyruvate: quinone oxidoreductase, transaminase B, L-lactate dehydrogenase and malate dehydrogenase, respectively. The deletions increased the availability of pyruvate for production of L-valine and L-alanine. In addition, the genes encoding enzymes of Ehrlich pathway were overexpressed, therefore, 2-keto acids were formed from pyruvate that could be transformed into butanol. The authors obtained the isobutanol yield up to 12.9 g/L [30]. This work can be further expanded in order to produce high amount of butanol and strain can get a new phenotype. Implementation of metabolic engineering along with synthetic biology for higher production at industrial scale with competitive prices can be possible in near future.

There are many limitation of bacterial production of butanol for industrial scale, such as toxicity, productivity, substrate cost and time. Additionally, butanol production by bacterial acetone-butanol-ethanol (ABE) fermentation faces some complications including product inhibition, slow growth rate, bacteriophage contamination and sporulation in solventogenesis phase. Moreover, it is hard to maintain strict anaerobic cultivation and two-stage multi-temperature fermentation reaction during the process. In order to overcome these issues of *Clostridium* fermentation, many attempts were made to prove *E. coli* as the potential host for butanol production, though some issues of bacterial fermentation still prevail [31,32].

3.4 Butanol Production via Yeast

Yeast is a eukaryotic microorganism which has better potential for industrial scale butanol production. Scientists have confirmed the butanol production from different wild-type yeast strains including *Saccharomyces cerevisiae, Kluyveromyces lactis* and *Pichia pastoris* [33]. Amongst these yeasts, a highest yield (270 mg/L) of 3-methyl-1-butanol was obtained from *K. lactis*. Thus, the study proved the ability of wild-type yeasts strain to produce butanol and provided an anticipation to engineer yeasts for biofuel production.

Considering the high n-butanol tolerance and phage resistance characteristic of *S. cerevisiae*, many efforts have been made to produce butanol using this system. Steen *et al.* [34] were the first to demonstrate biobutanol production in *S. cerevisiae*. Isozymes of the n-butanol producing pathway from different organisms were expressed in *S. cerevisiae*. Enzymes from *S. cerevisiae, E. coli, Clostridium beijerinckii*, and *Ralstonia eutropha* were tested for butanol production by substituting clostridial enzymes. The 3-hydroxybutyryl-CoA dehydrogenase from *C. beijerinckii* and acetoacetyl-CoA transferase from *S. cerevisiae* were found to be appropriate for butanol production (2.5 mg/L of butanol was obtained). A mutant *S. cerevisiae* strain with the potential to produce butanol by Ehrlich pathway was also reported [35]. Two genes encoding enzymes 2-keto acid decarboxylase (*KDC*) and alcohol dehydrogenase (*ADH*) were overexpressed. Simultaneously, *Ilv2*, which was concerned with catalysis of the first step in valine production pathway, was overexpressed. *PDC1* gene (pyruvate decarboxylase) was deleted to redirect the flux from ethanol production towards butanol. The butanol yield was increased by up to 13-fold and the titre value of wild-type and mutant strain was 11 mg/L and 143 mg/L, respectively [35].

In another study, Shi *et al.* [36] took advantage of an endogenous pathway involving threonine catabolism for enhancing the butanol production. The authors used previous strategies of overexpression of keto-acid decarboxylases (*KDC*) and alcohol dehydrogenase (*ADH*) and collaterally expressed *LEU1, LEU4, LEU2, LEU5, CimA, NFS1, ADH7* and *ARO10*. The final engineered strain produced up to 835 mg/L butanol which was 7-fold higher than the initial strain. *LEU1, LEU4, LEU2* and *LEU5* were responsible for the conversion of threonine to α-ketovalerate. NFS1 stabilized Leu1p which was involved in the formation of ketovalerate from threonine. CimA allowed for shortening of the endogenous pathway by forming citramalate from pyruvate and acetyl CoA, while conversion of ketovalerate to n-butanol required ADH7 and ARO10. Such attempts on *S. cerevisiae* provide proof for the utilization of heterologous organism for butanol production. *P. pastoris* has also faster growth rate and metabolism that can be further used for butanol production. Thus, these organisms have the potential to produce butanol with low input cost and high amount at competitive price.

3.5 Butanol Production using Microalgae and Cyanobacteria

Eukaryotic microalgae and cyanobacteria are distinct photosynthetic microorganisms having the capability to convert solar energy into carbon compounds. These organisms require sunlight, atmospheric

CO_2, water and trace amount of nutrients which allow conversion to the desired molecules. Algae are known to have more than 1,00,000 genetically diverse strains which allow it to stand out and develop promising biofuels. Since long, algae has caught the attention of researchers to divert it as the future source of biofuel production because of its ability to produce and accumulate lipids. In this light, biofuel production directly from CO_2 is an attractive approach as it can solve both energy and environmental problems. However, the productive potentials of cyanobacteria and algae remain largely unexplored. Even though, algae could be developed as biofuel factories due to their immense features as given below:

- Rapid biomass production
- High per-acre productivity
- Ability to produce and accumulate high amount of energy-rich oil (*Botryococcus* spp., *Nannochloropsis* sp., *Neochloris* sp.)
- Utilizes non-food based resources to meet its energy needs
- Flexibility and adaptability at varying conditions
- Multitude diversity
- Capability to grow on barren lands
- Eco-friendly as it accounts for > 40% of global carbon fixation [37]
- Wider portfolio of co-products
- Ease of genetic manipulation
- Ability to remove heavy metals and other nutrients from the waste streams.

Cyanobacteria or blue-green algae are the lone bacteria that possess chlorophyll a. Earlier, these were known to accumulate comparatively low levels of lipids. These lipid molecules were extracted and purified subsequently for acquiring energy-rich biofuels. Because of simple growth requirements, ability to endure adverse environmental conditions and easy genetic manipulation of its genome, these materials are a renewable alternative for biofuel production.

The photosynthetic cyanobacteria have already been engineered to produce biobutanol and its derivatives. *Synechocystis* sp. strain PCC 6803, a model cyanobacterium, exhibits plentiful variation in biomass and lipid production rates [38] under photoautotrophic as well as mixotrophic conditions, which make it a workhorse in context of cyanobacterium biofuel production. Isobutanol (a high energy dense compound) can be easily obtained from isobutyraldehyde through *S. elongatus* PCC 7942, by the addition of an alcohol dehydrogenase [39]. Moreover, it can also be synthesized by the addition of α-ketoisovalerate decarboxylase (*kivd*) from *L. lactis* and alcohol dehydrogenases (*adh*) in *Synechocystis* sp. PCC 6803 [40]. For maximizing the efficiency of cyanobacterial cell factories, oxygenic photosynthetic capability involving light and dark reactions of photosynthesis is needed to be optimized in different strains of cyanobacteria [41].

Metabolic engineering has assisted to introduce effective, productive and unique pathways. For instance, production of 1-butanol was successfully accomplished (14.5 mg/L) by transferring a modified CoA-dependent 1-butanol pathway [42]. In brief, it was achieved by transferring *hbd*, *crt*, and *adhE2* genes from *C. acetobutylicum*, *ter* gene from *Treponema denticola* and *atoB* gene from *E. coli* into wild-type *S. elongatus* PCC 7942 [42]. The production of 1-butanol has been further enhanced by 4-folds as compared to wild-type by inserting genes such as *AdhE2*, *Bldh* and *YqhD* [43]. In the production of n-butanol through the CoA-dependent route, CoA-acylating aldehyde dehydrogenase appears as hindering agent due to its oxygen sensitivity. This obstacle can be tackled by insertion of oxygen tolerant CoA-acylating aldehyde dehydrogenases (*PduP*) in the pathway of 1,2-propanediol degradation [44]. The ATP and co-factors are the driving forces that play a pivotal role to alter metabolic flux for the enhancement of butanol production [43]. Subsequently, isobutanol production in *Synechococcus* sp. strain PCC 6803 was successfully engineered to accumulate 90 mg/L from 50 mM bicarbonate in a shaking flask condition by addition of two heterologous genes *kivd* and *adhA* from the Ehrlich pathway without any inducer or antibiotics [38]. Additionally, in the presence of glucose source, up to 114 mg/L isobutanol could be produced. As the presence of high titres of isobutanol is toxic to the cells and has the probability to be degraded photochemically by hydroxyl radicals, it is obligatory to remove it by *in-situ* removal using oleyl alcohol as a solvent [38].

Another resolution of high titre butanol toxicity is to explore the regulatory mechanisms of butanol tolerance in cyanobacteria through transcription factor engineering. It could be resolved by the deletion of butanol-responsive two-component signal transduction systems (TCSTSs) genes, followed by deletion of the gene *slr1037*, an encoding novel response regulator in many different species of cyanobacteria [45]. It could be further confirmed by quantitative isobaric tags for relative and absolute quantification (iTRAQ) LC-MS/MS proteomics approach, which is a method based on tandem mass spectroscopy to determine the amount of proteins from different sources in quantitative proteomics. The reliability of iTRAQ LC-MS/MS could be confirmed with the real-time reverse transcription PCR technique [46]. Furthermore, 2-methyl-1-butanol (2MB) also drew attention of researchers as a potent energy-dense biofuel from CO_2. It can be produced by inclusion of citramalate pathway and enzymes such as ketoacid decarboxylase (*kivd*) and alcohol dehydrogenase (*yqhD*) in *S. elongatus* PCC7942 [47].

In the past decades, the advent of metabolic engineering, synthetic biology tools and CRISPR-Cas9 technique [9,48-53] have led to nascent era of sophisticated re-engineering of cellular machinery in a fast and predictable way. The CRISPR-Cas platform for gene editing in cyanobacteria is a modern concept that can result in yielding and enhancing biofuels [54]. In the present scenario, the cost of the biofuels is a major concern which could be diminished by reinforcing more research towards microorganisms for making them biofuels factories, especially by using cyanobacteria and algae because of its advantageous features which have come to light in the last two decades.

3.6 Conclusion and Future Remarks

It is indispensable to develop strain and production recovery technology for easy extraction of butanol. Butanol with prospects such as high energy and octane number as well as ability to be used directly in combustion engines proves as a fuel with high potential to replace gasoline, indicating better performance than ethanol. Implementation of microbial cell factories for mass production of biofuels seems to be a solitary method. Wild-type strains of certain microorganisms including bacteria, yeasts, algae and cyanobacteria have specific metabolic pathways for biosynthesis of butanol. For example, *C. acetobutylicum* is native producer of butanol. Apart from this, native butanol producer species and microbes from diverse genus are potent butanol producers. However, employing synthetic tools for performing metabolic engineering practices and deriving mutant strains with extraordinary ability to produce butanol is beneficial.

So far, the solventogenesis phase of *Clostridium* CoA pathway is engineered in several studies to obtain a strain aiming to reduce by-product formation, increase metabolic flux and butanol tolerance. The amino acid pathway is the second highly utilized pathway for butanol production in engineered strains. This pathway was engineered in *E. coli* and has shown maximum butanol production. Despite this, it is more convenient to utilize *S. cerevisiae* due to its industrial importance. As *S. cerevisiae* is more appropriate for present industrial infrastructure, more research focus has been devoted to this system. Furthermore, there are perspectives for the development and optimization of strains and diverse metabolic pathways to produce butanol. For example, generating a pathway in a strain with the ability to grow on a substrate to produce butanol would have a significant impact on biofuels industry. Metabolic engineering along with synthetic biology has gained more scientific and public interest due to simple design, building and testing processes. In near future, butanol production can be further improved and optimized by manipulating number of microorganisms that can grow on cheaper sugars or photosynthetic processes for industrial scale production.

Acknowledgements

This work was supported by The Puri Foundation for Education in India.

References

1. CO_2 Emission From Fuel Combustion: Overview (2017). *International Energy Agency*. Online: http://www.iea.org/publications/freepublications/publication/CO2EmissionsFromFuelCombustion2017Overview.pdf (accessed 21st May 2018).
2. Mohr, S. H., Wang, J., Ellem, G., Ward, J., and Giurco, D. (2015) Projection of world fossil fuels by country. *Fuel*, **141**, 120-135.
3. Shafiee, S., and Topal, E. (2009) When will fossil fuel reserves be diminished?. *Energy Policy*, **37**(1), 181-189.
4. Biodiesel Handling and Use Guidelines (BHUG) (2016). *Energy Efficiency and National Renewable Energy Laboratory, U.S. Department of Energy*, USA. Online: http://biodiesel.org/docs/using-hotline/nrel-handling-and-use.pdf (accessed 21st May 2018).
5. Coronado, C. R., de Carvalho, Jr., J. A., and Silveira, J. L. (2009) Biodiesel CO_2 emissions: A comparison with the main fuels in the Brazilian market. *Fuel Processing Technology*, **90**(2), 204-211.

6. Durre, P. (2007) Biobutanol: an attractive biofuel. *Biotechnology Journal*, **2**(12), 1525-1534.
7. Xue, C., Zhao, X. Q., Liu, C. G., Chen, L. J., and Bai, F. W. (2013) Prospective and development of butanol as an advanced biofuel. *Biotechnology Advances*, **31**(8), 1575-1584.
8. Alper, H., and Stephanopoulos, G. (2009) Engineering for biofuels: exploiting innate microbial capacity or importing biosynthetic potential?. *Nature Reviews Microbiology*, **7**(10), 715.
9. Singh, V. (2014) Recent advancements in synthetic biology: current status and challenges. *Gene*, **535**(1), 1-11.
10. Lee, S. K., Chou, H., Ham, T. S., Lee, T. S., and Keasling, J. D. (2008) Metabolic engineering of microorganisms for biofuels production: from bugs to synthetic biology to fuels. *Current Opinion in Biotechnology*, **19**(6), 556-563.
11. Singh, V., Mani, I., Chaudhary, D. K., and Dhar, P. K. (2014) Metabolic engineering of biosynthetic pathway for production of renewable biofuels. *Applied Biochemistry and Biotechnology*, **172**(3), 1158-1171.
12. Singh V., Chaudhary D. K., Mani I., Dhar P. K. (2016) Recent advances and challenges of the use of cyanobacteria towards the production of biofuels. *Renewable and Sustainable Energy Reviews*, **60**: 1-10.
13. Gohil, N., Panchasara, H., Patel, S., Ramirez-Garcia, R., and Singh, V. (2017) Book review: Recent advances in yeast metabolic engineering. *Frontiers in Bioengineering and Biotechnology*, **5**, 71.
14. Desai, R. P., Harris, L. M., Welker, N. E., and Papoutsakis, E. T. (1999) Metabolic flux analysis elucidates the importance of the acid-formation pathways in regulating solvent production by *Clostridium acetobutylicum*. *Metabolic Engineering*, **1**(3), 206-213.
15. Harris, L. M., Desai, R. P., Welker, N. E., and Papoutsakis, E. T. (2000) Characterization of recombinant strains of the *Clostridium acetobutylicum* butyrate kinase inactivation mutant: need for new phenomenological models for solventogenesis and butanol inhibition? *Biotechnology and Bioengineering*, **67**(1), 1-11.
16. Jang, Y. S., Lee, J. Y., Lee, J., Park, J. H., Im, J. A., Eom, M. H., Lee, J., Lee, S. H., Song, H., Cho, J. H., and Lee, S. Y. (2012) Enhanced butanol production obtained by reinforcing the direct butanol-forming route in *Clostridium acetobutylicum*. *MBio*, **3**(5), e00314-12.
17. Nair, R. V., Green, E. M., Watson, D. E., Bennett, G. N., and Papoutsakis, E. T. (1999) Regulation of the sol locus genes for butanol and acetone formation in *Clostridium acetobutylicum* ATCC 824 by a putative transcriptional repressor. *Journal of Bacteriology*, **181**(1), 319-330.
18. Tummala, S. B., Welker, N. E., and Papoutsakis, E. T. (2003) Design of antisense RNA constructs for downregulation of the acetone

formation pathway of *Clostridium acetobutylicum. Journal of Bacteriology*, **185**(6), 1923-1934.

19. Jiang, Y., Xu, C., Dong, F., Yang, Y., Jiang, W., and Yang, S. (2009) Disruption of the acetoacetate decarboxylase gene in solvent-producing *Clostridium acetobutylicum* increases the butanol ratio. *Metabolic Engineering*, **11**(4-5), 284-291.

20. Sillers, R., Chow, A., Tracy, B., and Papoutsakis, E. T. (2008) Metabolic engineering of the non-sporulating, non-solventogenic *Clostridium acetobutylicum* strain M5 to produce butanol without acetone demonstrate the robustness of the acid-formation pathways and the importance of the electron balance. *Metabolic Engineering*, **10**(6), 321-332.

21. Yu, M., Zhang, Y., Tang, I. C., and Yang, S. T. (2011) Metabolic engineering of *Clostridium tyrobutyricum* for n-butanol production. *Metabolic Engineering*, **13**(4), 373-382.

22. Schwarz, K. M., Grosse-Honebrink, A., Derecka, K., Rotta, C., Zhang, Y., and Minton, N. P. (2017) Towards improved butanol production through targeted genetic modification of *Clostridium pasteurianum. Metabolic Engineering*, **40**, 124-137.

23. Zhu, L., Dong, H., Zhang, Y., and Li, Y. (2011) Engineering the robustness of *Clostridium acetobutylicum* by introducing glutathione biosynthetic capability. *Metabolic Engineering*, **13**(4), 426-434.

24. Baez, A., Cho, K. M., and Liao, J. C. (2011) High-flux isobutanol production using engineered *Escherichia coli*: a bioreactor study with in situ product removal. *Applied Microbiology and Biotechnology*, **90**(5), 1681-1690.

25. Atsumi, S., Cann, A. F., Connor, M. R., Shen, C. R., Smith, K. M., Brynildsen, M. P., and Liao, J. C. (2008a) Metabolic engineering of *Escherichia coli* for 1-butanol production. *Metabolic Engineering*, **10**(6), 305-311.

26. Nielsen, D. R., Leonard, E., Yoon, S. H., Tseng, H. C., Yuan, C., and Prather, K. L. J. (2009) Engineering alternative butanol production platforms in heterologous bacteria. *Metabolic Engineering*, **11**(4-5), 262-273.

27. Shen, C. R., Lan, E. I., Dekishima, Y., Baez, A., Cho, K. M., and Liao, J. C. (2011) Driving forces enable high-titre anaerobic 1-butanol synthesis in *Escherichia coli. Applied and Environmental Microbiology*, **77**(9), 2905-2915.

28. Atsumi, S., Hanai, T., and Liao, J. C. (2008b) Non-fermentative pathways for synthesis of branched-chain higher alcohols as biofuels. *Nature*, **451**(7174), 86.

29. Li, S., Wen, J., and Jia, X. (2011) Engineering *Bacillus subtilis* for isobutanol production by heterologous Ehrlich pathway construction and the biosynthetic 2-ketoisovalerate precursor pathway overexpression. *Applied Microbiology and Biotechnology*, **91**(3),

577-589.

30. Blombach, B., Riester, T., Wieschalka, S., Ziert, C., Youn, J. W., Wendisch, V. F., and Eikmanns, B. J. (2011) *Corynebacterium glutamicum* tailored for efficient isobutanol production. *Applied and Environmental Microbiology*, **77**, 3300-3310.

31. Schadeweg, V., and Boles, E. (2016) n-Butanol production in *Saccharomyces cerevisiae* is limited by the availability of coenzyme A and cytosolic acetyl-CoA. *Biotechnology for Biofuels*, **9**(1), 44.

32. Swidah, R., Wang, H., Reid, P. J., Ahmed, H. Z., Pisanelli, A. M., Persaud, K. C., Grant, C. M., and Ashe, M. P. (2015) Butanol production in *S. cerevisiae* via a synthetic ABE pathway is enhanced by specific metabolic engineering and butanol resistance. *Biotechnology for Biofuels*, **8**(1), 97.

33. Azah, R. N., Roshanida, A. R., and Norzita, N. (2014) Production of 3-methyl-1-butanol by yeast wild strain. *International Journal of Biological, Biomolecular, Agricultural, Food and Biotechnological Engineering*, **8**(4), 412-415.

34. Steen, E. J., Chan, R., Prasad, N., Myers, S., Petzold, C. J., Redding, A., Ouellet, M., and Keasling, J. D. (2008) Metabolic engineering of *Saccharomyces cerevisiae* for the production of n-butanol. *Microbial Cell Factories*, **7**(1), 36.

35. Kondo, T., Tezuka, H., Ishii, J., Matsuda, F., Ogino, C., and Kondo, A. (2012) Genetic engineering to enhance the Ehrlich pathway and alter carbon flux for increased isobutanol production from glucose by *Saccharomyces cerevisiae*. *Journal of Biotechnology*, **159**(1), 32-37.

36. Shi, S., Si, T., Liu, Z., Zhang, H., Ang, E. L., and Zhao, H. (2016) Metabolic engineering of a synergistic pathway for n-butanol production in *Saccharomyces cerevisiae*. *Scientific Reports*, **6**, 25675.

37. Parker, M. S., Mock, T., and Armbrust, E. V. (2008) Genomic insights into marine microalgae. *Annual Review of Genetics*, **42**, 619-645.

38. Varman, A. M., Xiao, Y., Pakrasi, H. B., and Tang, Y. J. (2013) Metabolic engineering of *Synechocystis* sp. strain PCC 6803 for isobutanol production. *Applied and Environmental Microbiology*, **79**(3), 908-914.

39. Atsumi, S., Higashide, W., and Liao, J. C. (2009) Direct photosynthetic recycling of carbon dioxide to isobutyraldehyde. *Nature Biotechnology*, **27**(12), 1177.

40. Miao, R., Liu, X., Englund, E., Lindberg, P., and Lindblad, P. (2017) Isobutanol production in *Synechocystis* PCC 6803 using heterologous and endogenous alcohol dehydrogenases. *Metabolic Engineering Communications*, **5**, 45-53.

41. Angermayr, S. A., and Hellingwerf, K. J. (2013) On the use of metabolic control analysis in the optimization of cyanobacterial biosol-

ar cell factories. *The Journal of Physical Chemistry B*, **117**(38), 11169-11175.

42. Lan, E. I., and Liao, J. C. (2011) Metabolic engineering of cyanobacteria for 1-butanol production from carbon dioxide. *Metabolic Engineering*, **13**(4), 353-363.

43. Lan, E. I., and Liao, J. C. (2012) ATP drives direct photosynthetic production of 1-butanol in cyanobacteria. *Proceedings of the National Academy of Sciences USA*, **109**(16), 6018-6023.

44. Lan, E. I., Ro, S. Y., and Liao, J. C. (2013) Oxygen-tolerant coenzyme A-acylating aldehyde dehydrogenase facilitates efficient photosynthetic n-butanol biosynthesis in cyanobacteria. *Energy and Environmental Science*, **6**(9), 2672-2681.

45. Chen, L., Wu, L., Wang, J., and Zhang, W. (2014) Butanol tolerance regulated by a two-component response regulator Slr1037 in photosynthetic *Synechocystis* sp. PCC 6803. *Biotechnology for Biofuels*, **7**(1), 89.

46. Li, X., Wang, Q., Gao, Y., Qi, X., Wang, Y., Gao, H., Gao, Y., and Wang, X. (2015) Quantitative iTRAQ LC-MS/MS proteomics reveals the proteome profiles of DF-1 cells after infection with subgroup J Avian leukosis virus. *BioMed Research International*, **2015**, 395307.

47. Shen, C. R., and Liao, J. C. (2012) Photosynthetic production of 2-methyl-1-butanol from CO_2 in cyanobacterium *Synechococcus elongatus* PCC7942 and characterization of the native acetohydroxyacid synthase. *Energy and Environmental Science*, **5**(11), 9574-9583.

48. Stephanopoulos, G. (2012) Synthetic biology and metabolic engineering. *ACS Synthetic Biology*, **1**(11), 514-525.

49. Jinek, M., Chylinski, K., Fonfara, I., Hauer, M., Doudna, J. A., and Charpentier, E. (2012) A programmable dual-RNA–guided DNA endonuclease in adaptive bacterial immunity. *Science*, **337**(6096), 816-821.

50. Jinek, M., East, A., Cheng, A., Lin, S., Ma, E., and Doudna, J. (2013) RNA-programmed genome editing in human cells. *elife*, **2**, e00471.

51. Mali, P., Yang, L., Esvelt, K. M., Aach, J., Guell, M., DiCarlo, J. E., Norville, J. E., and Church, G. M. (2013) RNA-guided human genome engineering via Cas9. *Science*, **339**(6121), 823-826.

52. Singh, V., Braddick, D., and Dhar, P. K. (2017) Exploring the potential of genome editing CRISPR-Cas9 technology. *Gene*, **599**, 1-18.

53. Singh, V., Gohil, N., Ramirez Garcia, R., Braddick, D., and Fofie, C. K. (2018) Recent advances in CRISPR-Cas9 genome editing technology for biological and biomedical investigations. *Journal of Cellular Biochemistry*, **119**(1), 81-94.

54. Naduthodi, M. I. S., Barbosa, M. J., and van der Oost, J. (2018) Progress of CRISPR-Cas based genome editing in photosynthetic microbes. *Biotechnology Journal*, doi:10.1002/biot.201700591.

Chapter 4

Biofuels Production using Supercritical Water Gasification of Biomass

Sanjeet Mehariya,[1,2] Angela Iovine,[1,2] Patrizia Casella,[1] Tiziana Marino,[3] Simeone Chianese[2] and Antonio Molino[1,*]
1ENEA, Italian National Agency for New Technologies, Energy and Sustainable Economic Development, Department of Sustainability - CR Portici. P. Enrico Fermi, 1, 80055 Portici (NA), Italy
2Department of Civil and Building Engineering, Design and Environment, Università degli Studi della Campania "L.Vanvitelli", Real Casa dell'Annunziata, Via Roma 9, 81031 Aversa (CE), Italy
3CNR, Institute on Membrane Technology, National Research Council. Via Pietro Bucci, Cubo 17C, c/o University of Calabria, 87036 Rende (CS), Italy
Corresponding author: antonio.molino@enea.it

4.1 Introduction

The pragmatic climate change, energy crisis and resource scarcity as well as pollution will be the major issues in near future. The growing demand of energy could be substituted with renewable energy sources, which subsequently minimize the pollution via resource recycling. The renewable energy can lead to zero emission of greenhouse gases (GHG) in comparison to conventional fossil fuels [1,2]. The renewable energy could be derived from various biomass sources such as municipal solid wastes (MSWs) and other organic wet waste from food and agricultural sectors. The recent estimation by Food and Agricultural Organization (FAO) indicates that ~$750 billion worth food waste (FW) is generated worldwide every year [3]. Globally FW is accountable for ~20 million tons of CO_2 equivalent GHG emissions per year [4,5], thus, demanding eco-friendly approach to overcome these issues. Wet biomass with different moisture contents could be used as feedstock for several technologies for biofuel production with different energy efficiencies, i.e., anaerobic digestion, liquefaction, pyrolysis, supercritical water gasification (SCWG) and thermal gasification. Moreover, the large portion of organic wastes are wet biomass, which have water content of 50%, frequently up to above 90% of dry weight (DW) [6-9]. Various bio-

Biofuels, edited by Vikas Mittal
© 2018 Central West Publishing, Australia

mass materials have been categorized in Table 4.1, which originate from diverse sources and have different moisture content. Also, the conversion efficiency of biomass depends on moisture content using

Table 4.1 Moisture and ash content of various biomass from different sectors

Sector	Physical state/type	Properties	
		Moisture (%)	Ash (%)
Agriculture	Dry	15-50	2.2-17
	Wet	74-92	27-35
	Dry	75	17-28
Forest	Dry	40-60	0.4-5
	Dry	40-50	0.4-5
Industry	Dry	10-30	0.71-18.34
	Wet	90	3.8-5.9
	Black liquor	90	3.8-5.9
Waste	Biodegradable waste	30	36
	Demolition wood	30-40	0.58
	Sewage sludge	70-80	26.4
	Dry	35	39.4
	Wet	75-80	8.4

different available technologies, as shown in Figure 4.1. Schematic of the transformation of biomass to fuels is also shown in Figure 4.2 [2,10,11]. Generally, dry wood biomass used for thermochemical conversion, i.e., pyrolysis or gasification technologies, requires a water content below 10% of DW. However, wet biomass is not considered as favorable feedstock for thermochemical conversion due to the high cost of drying, which could result in negative energy recovery [12]. Leible *et al.* [13] reported that the net energy recovery of raw sewage was -1700 kJ/kg wet basis. However, SCWG allows to overcome the obstacle to convert the wet biomass into syngas or liquid biofuels at the critical point of water (T = 374 °C and P = 22.1 MPa) due to its endothermic nature of reaction [11,14-18]. The SCWG technology has higher energy efficiency as compared to other technologies for wet biomass, which has the water content up to 90% of DW [19]. The SCWG process depends on the unique properties of water near the critical or supercritical thermodynamic state. During the supercritical thermodynamic state, the number of hydrogen bonds is lower, which sharply lowers the dielectric constant

☐ Moisture content in feed   ☐ Moisture content in feed
▨ Moisture content in feed   ■ Moisture content in feed

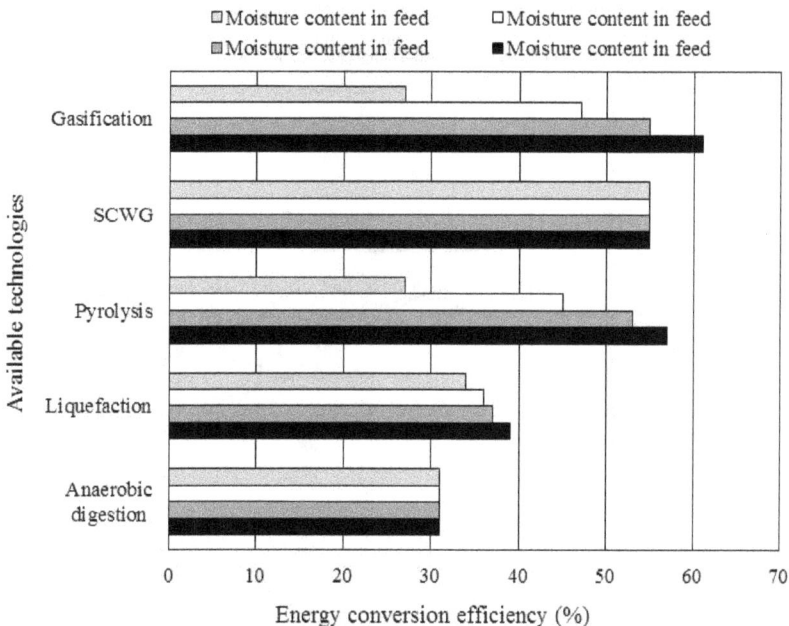

Figure 4.1 Comparison of energy conversion efficiency of different available technologies for biomass conversion.

of water due to weaker hydrogen bonds. Therefore, supercritical water (SCW) acts as solvent for organic components and all undesirable compounds contained in the syngas (e.g. tars) are solubilized in the liquid phase [20-23].

SCWG is a process in which biomass is converted to gaseous products (mainly composed of H_2, CH_4, CO and CO_2) with the main aim of producing highly flammable gas (H_2, CO and CH_4). The gasification of biomass at SCW is a complex process, which involves the formation of organic solvents through biomass hydrolysis and further decomposition of organic compounds to gaseous products. Theoretically, the simplified overall SCWG reaction is described in eq. (1), where glucose is used as model compound as biomass for complete conversion to hydrogen [23].

$$C_6H_{12}O_6 + 6H_2O \rightarrow 6CO_2 + 12H_2 \dots\dots\dots\dots\dots\dots\dots\dots\dots\dots\dots (1)$$

During biomass gasification at SCW, different chemical reaction pathways take place including biomass decomposition, steam reforming, water - gas shift, methanation reactions and the Boudouard

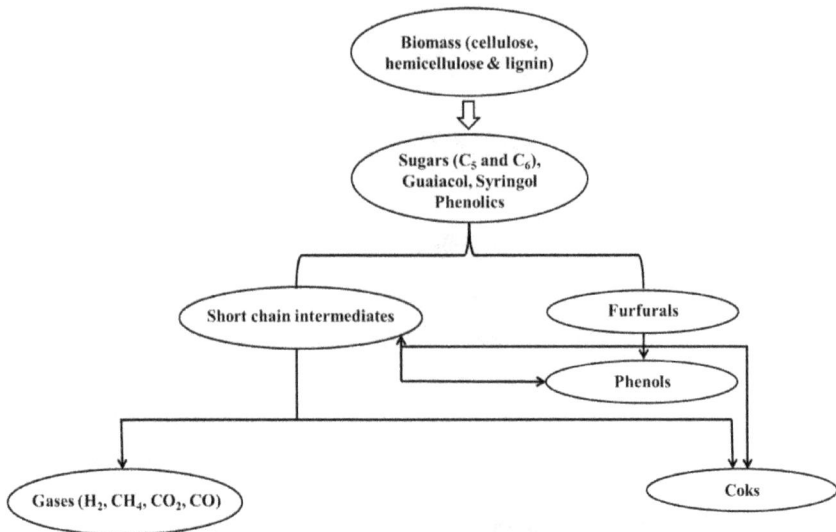

Figure 4.2 Flow diagram of schematic transformation of biomass to fuels.

equilibrium. Primarily, carbon monoxide is formed, which reacts with water to carbon dioxide and hydrogen as described in Eq. (2) [15,24,25].

$$CO + H_2O \leftrightharpoons CO_2 + H_2 \dots\dots\dots\dots\dots\dots\dots\dots\dots\dots\dots\dots\dots\dots . (2)$$

At temperatures near to the critical point of water, methane is the preferred product as described by eq. (3) and thermodynamically favored at lower temperatures with higher dry mass content of biomass [7,23].

$$C_6H_{12}O_6 \rightarrow 3CH_4 + 3CO_2 \dots\dots\dots\dots\dots\dots\dots\dots\dots\dots\dots . (3)$$

At temperatures at critical point, water hydrolysis and CO formation reaction accrues as shown in eq. (4), with typical hydrogenation catalysts like nickel or other noble metals [26].

$$CO + 3H_2 \leftrightharpoons CH_4 + H_2O \dots\dots\dots\dots\dots\dots\dots\dots\dots\dots\dots . (4)$$

Considering the overall reactions and equilibrium involved in the CH_4 and H_2 production, it has been observed that a few main factors could influence biomass gasification at SCW. Primarily, these include biomass to moisture content ratio, catalysts for promoting reaction rate and to avoid unwanted products as well as temperature, where higher temperatures favor hydrogen and lower methane production. Hydrogen is a promising source of energy, which has a crucial role for the development of a bio-economy. H_2 has a wide range of uses in refineries for treatment of heavy crude oil and hydrocarbons production from hydrogenation process, upgrading bio-oils, synthesis of liquid transportation fuels, etc. [20,26]. Gasification of biomass at SCW has great potential with regards to the optimization of chemical syntheses in comparison with conventional gasification, pyrolysis and electrolysis [27]. However, biomass consists of many different components and it is a complex task to understand their influence on SCWG. This chapter covers some of the key factors to contextualize the fundamentals of biomass SCWG and knowledge is updated by incorporating recent literature.

4.2 Supercritical Water Gasification of Biomass Components

Lignocellulose biomass is the non-edible part of the plant, which contributes up to 90% of the available land biomass. It is known as residue or waste, therefore, it is a multipurpose feedstock to produce numerous bio-products, such as bio-based chemicals, biofuels and biopolymers [23,28,29]. The lignocellulosic biomass can be derived from various sources, such as agricultural residues, forest residues and portions of municipal solid waste (MSW) with an estimated annual production of 200 billion tons [30,31]. The main composition of lignocellulose are cellulose, hemicellulose and lignin [32], the normal composition of lignocellulosic biomass consists of 40-55 wt% cellulose fibers and 15-35 wt% hemicellulose embedded in 20-40 wt% of lignin. Therefore, the biological conversion for the degradation is restricted due to their recalcitrant nature. The majority of microorganisms are incapable to break-down lignin, which is a significant portion of the biomass [33]. The break-down of lignin from the sugar-rich biomass requires a pretreatment process, which is energy intensive and expensive [14,34]. Also, the variable amounts of moisture, inorganic ash and other constituents such as proteins or resins are contained in lignocellulosic biomass [10]. Therefore, SCWG is an ideal technology for H_2 production, with biomass composition a determining parameter. Moreover, the amount of ash, protein and lignin could change the conversion efficiency towards H_2. Also, the biomass composition, operating conditions and the presence of catalyst could result in various concentrations of H_2. The impact of these factors on the final product distribution is presented in the following sections.

4.2.1 Cellulose and Hemicellulose

Cellulose is synthesized in the plant cell membrane, which is glucose polymer consisting of β (1,4) linked D-glucose subunits which are aggregated by hydrogen bonding and van der Waals forces [35]. The linear nature of cellulose particles is accountable for developing intra and inter-molecular hydrogen bonding [36]. Cellulose contributes the major proportion in biomass composition and its hydrolysis results in the formation of glucose. Glucose is a monomer of cellulose and is the main product of cellulose decomposition in SCWG [37]. Also, hemicellulose is a combination of polysaccharides composed of C_5 and C_6 sugars such as arabinose, glucose, mannose and xylose along with sugar acids such as methylglucuronic and galaturonic acids [38]. The conversion of cellulose and hemicellulose for gas products has been commonly studied with model substances using pure cellulose and xylan. Also, SCWG of cellulose results in higher H_2 production as compared to hemicellulose, probably due to its intrinsic larger H_2 content [39,40]. Moreover, hydrolysis of glucose near and above the critical temperatures [41] as well as other chemical reactions have been described for SCWG of glucose [7,42-45]. The gas products of glucose SCWG are produced at higher temperatures and majorly consist of H_2, CO, CO_2, CH_4 and traces of hydrocarbons [46]. However, SCWG of cellulose alone leads to lower H_2 concentration in gas product mixture and leads to minimization of the gasification efficiencies to 50%. On mixing cellulose with a small amount of sodium carboxymethyl cellulose and a Ru/C catalyst, the efficiency improves significantly and the quantity of H_2 concentration in gas mixture increases up to four-folds [47]. Comparable results were obtained in other studies, where nickel-based catalysts exhibited greater activity and higher H_2 selectivity as compared to ruthenium-based catalysts [48,49]. Particularly $Ni/\alpha\text{-}Al_2O_3$ and Ni/hydrotalcite led to higher H_2 concentration in the mixture of gas products. Also, the studies indicated that the amount of H_2 gas produced to the amount of H_2 available in the feed was approximately same for cellulose and hemicellulose. Therefore, the composition of cellulose and hemicellulose is an important parameter for achieving higher selectivity of H_2. Also, the catalysts play a key role during the decomposition of biomass at different subcritical and supercritical conditions. The reaction mechanism of glucose gasification at high pressure is demonstrated in Figure 4.3.

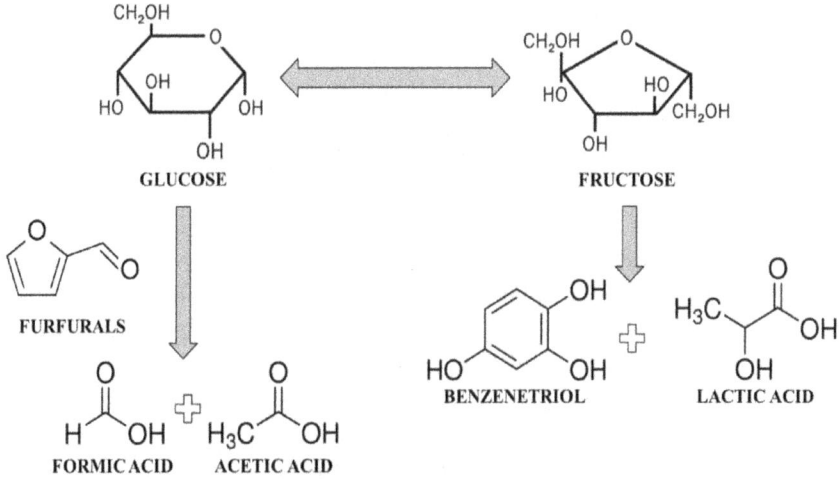

Figure 4.3 Representation of the reaction pathway of glucose gasification at high pressure.

4.2.2 Lignin

Lignin is a high molecular-weight compound with a complex structure, which acts as a binding agent that grasps cellulose and hemicellulose together and provides firmness to the lignocellulosic network. The three monomers of lignin composed of p-coumaryl alcohol, coniferyl alcohol and sinapyl alcohol act as building blocks, which provide resistance to the biomass against chemicals and enzymes [38]. It is a highly branched phenyl propane polymer linked with C-O-C and C-C bonds, which contains phenolic, methoxyl, hydroxyl and terminal aldehyde groups in the side chain [50,51]. It is challenging to obtain cellulose and hemicellulose from biomass in the presence of lignin. The complex structure of lignin leads to lower conversion degree in high value-added products [52,53]. The decomposition mechanism of lignin occurs at near critical conditions with short residence time of around 24 seconds [54]. SCWG initiates the synthesis of H+ and OH- ions to break-down lignin into reactive low molecular weight fractions, such as phenolic compounds (e.g., cathecol, guaiacol and phenol) and formaldehydes [23,55,56]. The lignin SCWG results mainly in four phases, i.e., aqueous (e.g., alcohols, aldehydes, catechol and phenols), gas (e.g., H_2, CO, CO_2 and CH_4), oil (e.g., phenolics, PAHs and heavy hydrocarbons) and solid phase (e.g., char) [57-59]. The decomposition pathway of phenols during SCWG is comparatively different from glucose. Furthermore, these products polymerize into higher molecular weight compounds via condensation/cross-linking reactions, leading to char formation [60-63]. The monomers and oligomers have a propensity to repolymerize with aldehydes to form solid residues (phenolic char) at high temperature (>400 °C) with long residence time of around 60 minutes [64]. Addition of phenol to lignin significantly inhibits the polymerization chain reaction, thereby, leading to the formation of lignin like residue in SCWG of lignin [65]. Generally, SCWG of lignin favors CH_4 and CO_2, while it has been stated that high temperatures are essential to complete the reaction with maximum H_2 selectivity [25].

Huelsman *et al.* [66,67] carried out studies to understand the decomposition mechanism of phenol and guaiacol in SCWG at different temperatures, with and without catalysts. The results indicated that CH_4 and H_2 rich gas could be obtained at higher temperatures (700 °C) and the addition of Ni catalyst significantly improved H_2 production. Additionally, the decomposition of phenols resulted in numerous intermediate compounds such as dibenzofuran and other dimers of phenol that act as precursors to char formation. The authors also established a direct relationship between the concentrations of phenol, water (in terms of water density) and gas composition. Also, the relationship of water density (phenol, water and the gas composition) within the hydrothermal media was observed to be a key parameter. Several studies have reported that higher water densities increase the rate of lignin decomposition for the production of oils and gases, probably by improved hydrolysis with higher water density. In summary, it can be concluded that lignin splitting is supported by active H_2 that could be derived from cellulose in biomass and could decrease the H_2 yield, if the gasification reaction is incomplete [23,51].

4.2.3 Proteins and Lipids

The amino acids are the building blocks of proteins, therefore, the hydrolysis of proteins at supercritical conditions is a comparatively slow process due to the stability of the peptide bond. The peptide bond links amino acids together into proteins with C–N bond between the carboxyl and amine groups present in all amino acids. The conversion rate has the relationship with type of amino acid used for the hydrothermal hydrolysis. However, the reaction pathway occurs in two steps, first is the formation of carbonic acid and amines by decarboxylation pathways, while second is the formation of ammonia and organic acids by deamination pathway [68,69]. A recent study carried out the decomposition of carbohydrates rich biomass and proteins rich biomass at supercritical conditions [70]. The protein rich biomass resulted in lower gas production together with severe corrosion of the reactor. The results indicated that lower gas yield resulted due to the inhibition of decarboxylation and decarbonylation reactions [70].

Lipids are non-polar compounds with molecular structures comparable to hydrocarbon fuels and can be used as alternatives for conventional hydrocarbons. Only a few studies have reported SCWG of lipids for generating free fatty acids via hydrolysis at temperatures below the critical point of water. The hydrolysis reactions of lipids and water are significantly influenced by their phase behavior. In a related study, SCWG was carried out using coconut, linseed and soybean oils resulting in a mixture of saturated and unsaturated fatty acids. Hydrolysis reaction occurred within the first 15-20 min, yielding conversion of 97% or higher. Afterwards, the hydrolysis products underwent decarboxylation and decarbonylation reactions to produce CO and CO_2, and the olefins decomposed further into H_2. Also, the fatty acids started to thermally degrade via decomposition, pyrolysis or polymerization at supercritical point [71].

4.3 Role of Operating Conditions

The operating conditions describe the potential of the reaction driving forces, i.e., thermodynamics and kinetics. These can influence product selectivity and rate of gasification efficiency at supercritical conditions. To underline these effects, the impact of different operating conditions has been discussed in following sub-sections.

4.3.1 Role of Catalysts

This section discusses the influence of various catalysts on SCWG reaction, which play key role in biomass transformation into biofuels. Several researchers have investigated the effect of different catalysts during SCWG reaction, such as alkali or heterogeneous metals or carbon catalysts such as charcoal/activated carbon (AC) [24,49]. All of these materials exhibit significant catalytic effect on the SCWG reaction with respect to product selectivity as well as operative conditions by reducing the operating cost. The capability to chop C-C bonds and reduce activity towards C-O bond cleavage during the biomass gasification allows to be a preferred catalyst in SCWG. Furthermore, implication of catalysts minimize the prerequisite of extreme operating conditions and decrease char and tar formation in SCWG [72]. However, catalysts could also drop the conversion rate, which depends on properties of the reactor used [73,74]. Alkali catalysts are miscible in the water and are more effective for SCWG of biomass, however, their re-use and recovery are challenging. Alkali metal base catalysts such as K_2CO_3, $KHCO_3$, KOH, Na_2CO_3 and NaOH have been reported to be effective catalysts for SCWG. Alkali catalysts have been observed to enhance the WGS reaction and H_2 yields [24]. Lu *et al.* [75] observed that K_2CO_3 catalyst enhanced the H_2 yield up to two times as compared to the yield without catalyst at the same operating conditions. The addition of KOH increased the yields of H_2 and CO_2 during the SCWG of industrial inorganic waste and pyrocatechol, although CO yield decreased [72,76]. Onwudili *et al.* [77] investigated the catalytic effect of NaOH during the SCWG of glucose and other biomasses in a batch reactor which resulted in 80% increase in H_2 volume at 450 °C and 34 MPa. Yanik *et al.* [60] studied the influence of K_2CO_3, Trona, Na-$HCO_3 \cdot Na_2CO_3 \cdot 2H_2O$, red mud and Raney Ni on SCWG of corn stalk, corn cob and tannery wastes. It was observed that Trona was most effective among all the tested catalysts with maximum yields of H_2. Also, K_2CO_3 and Trona catalysts were tested for SCWG of acorn, cauliflower residue, extracted acorn, hazelnut shell and tomatoes residue at 600 °C and 35 MPa [78]. Trona enhanced the yields of H_2 and CO_2, while reduced the yield of CO by greater extent as compared to K_2CO_3. Sheikhdavoodi *et al.* [79] carried out a study for maximizing H_2 production using Raney nickel with alkali metals and AC as catalyst in SCWG of sugarcane bagasse. The results indicated that the catalyst was essential to achieve high H_2 yields with KOH and NaOH, yielding the highest amount with 6.6 and 5.6 mol/g respectively. Though Raney nickel exhibited the highest biomass conversion efficiency, however, the H_2 yield was 5-fold lower with respect to the yield achieved in the presence of alkali metals. Azadi and Farnoon

[49] suggested that Raney nickel catalysts strongly promoted the C-O bond cleavage and supported hydrogenation of CO, which resulted in a CH_4-rich gas mixture.

Heterogeneous metal catalysts have advantage over alkali catalyst in the sense that these are relatively easy to recover, which reduces the operating cost. Different transition metals such as Ni, Ru, Cu and Co exhibit high catalytic activity and effectiveness for SCWG of biomass. Ni-based catalysts are extensively used in SCWG of biomass due to high activity and performance as compared to noble metals [39,80-82]. Ni catalysts can catalyze the aqueous products of biomass to gases and support the hydrogenation reaction, which can result in greater CH_4 and CO_2 yield via methanation reaction. Also, Ni catalysts lead to a high carbon conversion efficiency, however, the H_2 production is lower due to the consumption for hydrogenation [79]. Various support and promoter metals for Ni catalysts have been studied to influence catalytic activity and stability. One way of boosting the hydrogen production while using a nickel catalyst is by modifying its surface, by variation of the support material. A study was carried out using Ni catalyst with support of γ-Al_2O_3 for SCWG of glucose at 600 °C and 24 MPa and H_2 yield of 38.4 mol/kg glucose was obtained [31]. Azadi *et al.* [48] compared the catalytic effect of Raney nickel, Ni/α-Al_2O_3 and Ni/hydrotalcite with Ru/C and Ru/γ-Al_2O_3 as well as their influence on H_2 selectivity and yield during SCWG of cellulose, lignin, bark and model carbohydrates. The results suggested that higher H_2 selectivity and yield could be achieved using Ni/α-Al_2O_3 and Ni/hydrotalcite. Additionally, Mg and Ru have also been reported to be effective promoters for Ni catalysts. Furusawa *et al.* [82] investigated the effect of Ni/MgO catalyst in SCWG of lignin and observed that H_2 yields increased from 14.1% to 46.2% when 0.05 g of 20 wt% Ni/MgO catalyst was added at 400 °C. Also, highest carbon recovery could be achieved, when calcined at 600 °C. In addition, other noble metals such as palladium, rhodium or ruthenium have exhibited superior catalytic activity and potential for biofuel production in SCWG of biomass [23]. Yamaguchi *et al.* [83] used palladium, ruthenium, platinum, rhodium and nickel as carbon supported catalysts during SCWG and observed higher H_2 yield in the following order Pd > Ru > Pt > Rh > Ni.

Several carbonaceous materials such as spruce wood charcoal, macadamia shell charcoal and coal activated carbon have also exhibited ability as catalysts as well as catalyst support materials. The high surface area and minor corrosion issues resulted in higher stability in SCWG. In addition, the low cost makes these materials as a good choice of catalysts in SCWG of biomass. Studies carried out using various carbonaceous materials to observe the effect on gas formation during the SCWG of biomass showed that the presence of these materials enhanced the carbon conversion rate, while reducing the CO content of the gaseous products [10,12]. In another study, Basu and Mettanant [10] suggested that carbon catalysts enhanced CH_4 and H_2 yields due to progress in WGS reaction. Activated carbon resulted in higher catalytic activity as compared to Raney nickel for H_2 yields, however, it was not as high as alkali metal catalysts [79].

In conclusion, catalysts play a key role in SCWG of biomass in terms of improving the product yield as well as selectivity. The effect of various catalysts and there support materials is summarized in Table 4.2 with respect to H_2 yield. It also generally suggested to reduce the reaction temperature, thus, improving the thermodynamic efficiency of the process.

4.3.2 Effect of Temperature

Reaction temperature has a key role in SCWG of biomass as it significantly influences the performance of SCWG process. It has a significant impact on the specific yield of gasification especially in the absence of catalyst. Lower temperature decreases the carbon conversion efficiency, therefore, reducing the H_2 yield, while higher temperature enhances the gasification rate and increases H_2 yield [25]. In fact, the gasification process can be divided into three groups depending on the primary product of SCWG [101]. Generally, high temperature SCWG (500-800 °C) is thermodynamically less efficient in respect to low temperature SCWG, because external energy could be needed to continue the process [10]. Study carried out by Matsumura *et al.* [102] to observe the effect of gasification temperature of biomass revealed that low temperature gasification promoted CH_4 selectivity, while high temperature gasification enhanced H_2 selectivity. Other research studies have also investigated the role of temperature on product yield in SCWG of both biomass and model biomass (glucose) [47,75,103]. Lu *et al.* [75] carried out the SCWG of wood sawdust and suggested that gasification efficiency increased and conversion efficiency was higher than 10% when temperature increased from 600 to 650 °C. Consequently, H_2, CO_2, and CH_4 yield also increased with higher temperature at 30 MPa. Other study carried by Jesus *et al.* [103] observed strong influence of temperature on the gas yield in SCWG. During the SCWG of corn starch, the gasification yield enhanced from 41% to 92% with increasing temperature from 550 to 700 °C, which resulted in a higher reduction in total carbon in effluent. In addition, the SCWG of clover grass also

Table 4.2 Effect of different catalysts and operating conditions on H_2 production

Feedstock	Reactor	Operating (T and P)	Catalyst	H_2 yields (mol/kg)	Ref.
Glucose	Batch	400 °C, 24.5 MPa	Without catalyst	1.3	[84]
			Ni/γ-Al_2O_3	10.5	
			Ni/CeO_2-γ-Al_2O_3	12.7	
		400 °C, 22.5-25 MPa	Ni-Mg-Al	7.2-11.8	[85]
		550 °C, 36 MPa	Ru/α-Al_2O_3	10.8	[86]
			Ru/α-Al_2O_3/NaOH	21.1	
			Ru/α-Al_2O_3/CaO	14.7	
Xylan				10.7	
Sawdust				10.4	
Cellulose				9.1	
		365-500 °C	Without catalyst	0.1-2.9	[87]
		500-600 °C	Ni, Cu and Fe	0.3-2.2	
Lignin		365-725 °C	Without catalyst	0.1-7.5	
Xylan		500-600 °C	Ni, Cu and Fe	0.9-2.7	
Sugarcane bagasse		400 °C	Without catalyst	0.5	[57]
			Ru/AC	1.92	
			Ru/TiO_2	3.2	
Sea weed		500 °C, 30.2-33.5 MPa	Without catalyst	11.8-16	[88]
Hog manure		500 °C, 31 MPa	Pd/AC, Ru/Al_2O_3, Ru/AC, AC,	0.4-0.5	[89]

			NaOH		
Glucose	Continuous	400 °C, 24.5 MPa	Without catalyst	1.3	[84]
			Ni/γ-Al$_2$O$_3$	10.5	
			Ni/CeO2-γ-Al$_2$O$_3$	12.7	
		550-700 °C, 28 MPa	Ni/AC, Ni-Y/AC, Ni-Fe/AC, Ni-Co/AC	6.1-38-3	[90]
		575-725 °C, 28 MPa	Without catalyst	20	[91]
			Ni/AC	30	
Biomass		575-725 °C, 28 MPa	Without catalyst	1.7	[92]
			Ni/γ-Al$_2$O$_3$	38.3	
		600 °C, 24 MPa	Ru-Ni/γ-Al$_2$O$_3$	25.5-54	[31]
		600-750 °C, 24 MPa	Ru-Ni/γ-Al$_2$O$_3$	29-54.8	
		600-767 °C, 25 MPa	Without catalyst	31.7-63.8	[93]
		700 °C, 24 MPa	Ru-Ni/γ-Al$_2$O$_3$, Ni/γ-Al$_2$O$_3$	50	[94]
		750-800 °C, 22-33 MPa	Without catalyst	16.6	[95]
		600-750 °C, 24 MPa	Without catalyst	6.8-11.3	[96]
		600 °C, 35 MPa	Without catalyst	4.9-20.2	[97]
			K$_2$CO$_3$	18.5-32	
			Trona	19-31.1	
		400-600 °C, 25 MPa	Without catalyst	13-26.3	[98]
Glucose	fluidized bed	550-650 °C, 23-27 MPa	Without catalyst	2.5-7.7	[99]
Corn cob			Without catalyst	2.6-12	
Sewage sludge		480-540 °C, 25 MPa,	KOH, K$_2$CO$_3$, NaOH, Na$_2$CO$_3$	1.9-15.5	[100]

exhibited a major role of temperature on product yield, as the gasification efficiency increased from 0.62 to 0.88, while the total carbon concentration of the liquid residue decreased from 0.11 to 0.03 mg, as temperature was increased from 625 to 700 °C [103]. Hao *et al.* [104] conducted the experiment at 25 MPa to observe the effect of temperature on SCWG of model biomass (glucose). Increase in SCWG reaction temperature from 500 to 650 °C enhanced the gasification and carbon efficiencies by 300% and 167%, respectively. Therefore, H_2 yield increased by 46%, while CO reduced by 74%. In summary, the above discussion suggests that the temperature in the range of 500-700 °C significantly improves the economics of SCWG process due to enhanced H_2 yield with increased temperature, whereas total organic carbon reduces in liquid phase products.

4.3.3 Effect of Pressure

Pressure is another thermodynamics parameter whose role on biomass gasification in SCW is relatively difficult to understand. The increasing pressure enhances density, dielectric constant and ionic products in SCWG of biomass [24]. Hao *et al.* [104] observed that increase in pressure enhanced the gasification efficiency as well as yield of H_2 and CO_2 at 500 °C in SCWG of model biomass. Basu and Mettanant [10] suggested that pressure has a evident effect on the SCWG of biomass at higher temperatures. Ji *et al.* [105] observed that increase in pressure from 15 to 27.5 MPa enhanced the gasification efficiency and H_2 and CH_4 yields, along with lower CO_2 yields, in SCWG of lignin. In contrast, few studies have reported that H_2 yield decreased with increasing pressure in SCWG of glucose at different temperatures (400-600 °C) [97]. Molino *et al.* [46] also reported that the concentration of liquid phase products dropped with increasing pressure from 110 bar to 320 bar in SCWG of zootechnical sludge at 310 °C. Thus, the role of pressure is quite complex, therefore, further research should be carried out to understand its overall influence on SCWG of biomass.

4.3.4 Effect of Feed Concentration

The feedstock of a SCW gasifier comprises mainly solid or semi-solid biomass with moisture content up to 90%, as reported in Table 4.1. The solid and moisture contents of biomass feedstock have significant role on gasification efficiency at SCW. Guo *et al.* [25] reported that higher feed concentration has a negative impact on SCWG reaction, because lower concentration of biomass leads to quick gasification, while it can be difficult with high concentration at same operating conditions. Therefore, it can minimize the viability of SCWG of biomass for industrial operations. The SCWG of model biomass decreased CH_4, CO_2 and H_2 yield, while CO yield enhanced, as glucose concentration in the feedstock increased. Byrd *et al.* [106] reported that increasing concentration of glucose from 1 to 5% decreased H_2 yield at 24.8 MPa and 700 °C. Another major challenge associated with higher biomass concentration is the pumping issue in the reactor during SCWG at higher pressure.

4.3.5 Effect of Reactor Design, Geometry and Reactor Material

Literature reports demonstrate that various reactor configurations have been employed during the SCWG of biomass. Commonly, batch and continuous reactors are used, where batch reactors work in two modes such as high pressure autoclaves and diamond anvil cells [107], while continuous reactors can be in the form of continuous flow reactors. Continuous flow reactors could be tubular reactors and continuous stirred tank reactors. In addition, fluidized bed reactors have emerged recently with continuous reactor configuration for SCWG of biomass. Batch reactors have a difficulty in splitting the different reactions occurring during the heating and cooling stages, therefore, reaction with short residence time is not possible to study. As compared to batch reactor, tubular flow reactors are suitable for reactions with short residence time, however, plugging is a major issue. Guo *et al.* [25] reported that plugging issue appear with high concentration of feedstock, due to char formation. As a consequence, plugging and mixing pose major hindrance to industrial application of SCWG. To overcome these obstacles, continuous stirred vessel reactors could be used with integration of tubular flow and autoclave reactors, therefore, biomass could be mixed and heated homogeneously in the reactor, though process needs additional energy for heating and is more complex [41,75,108]. Matsumura and Minowa [109] recommended bubbling fluidized bed reactor for continuous biomass gasification at SCW condition. Several studies reporting different reactors for SCWG of biomass and other model biomass compounds are summarized in Table 4.2. It also summarizes the performance of extremely different reactors, i.e., continuous and fluidized bed reactors for SCWG of biomass for H_2 production. Fig-

ure 4.4 also represents the schematic of a supercritical water gasification in a bench-scale reactor.

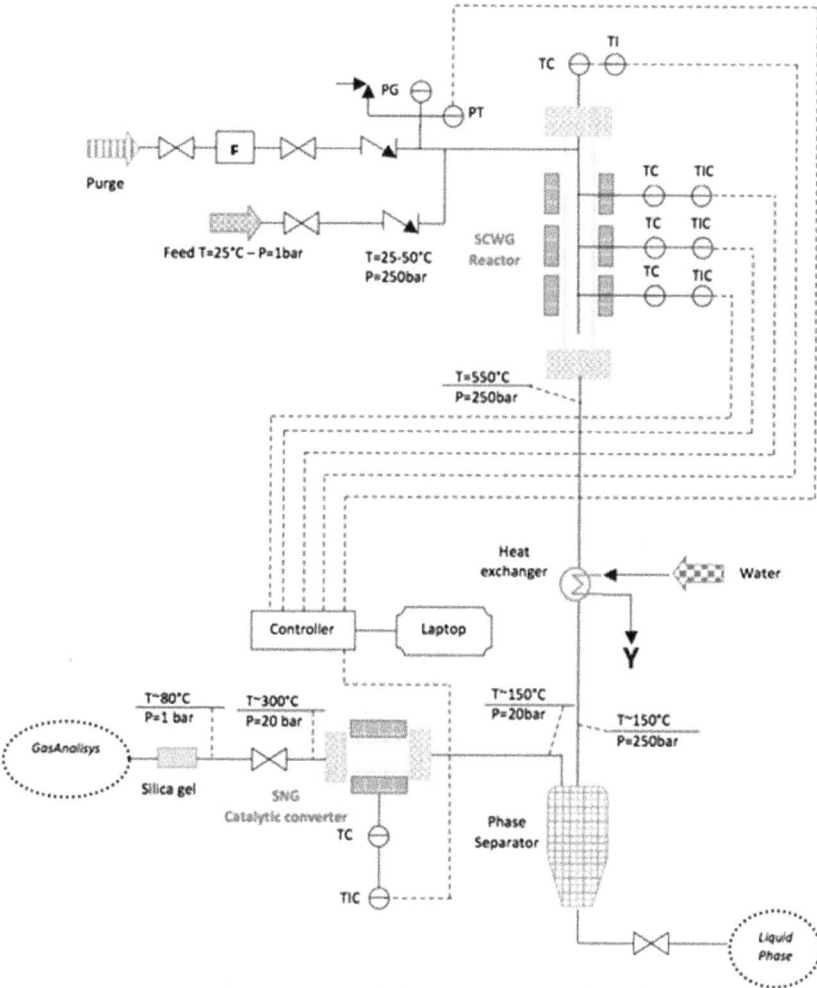

Figure 4.4 Combined SCGW + catalytic SNG converter experimental
apparatus. Reproduced from Reference 16 with permission from Elsevier.

Commonly, small reactors of few millimeter diameter are used
for SCWG during laboratory scale experiments. During the scale up,
reactors with diameters in the meter scale are used to observe the
effect of hydrodynamics, mass transfer and heat transfer inside the
reactors, as it is not possible in small scale reactors. Therefore, this
aspect needs greater attention to study the thermodynamics and ki-

netics of SCWG technology. Two different sizes of reactors were used for comparison by Hao *et al.* [104] and Lu *et al.* [75]. Studies were carried out to observe the effect of reactor inner diameter of 6 mm and 9 mm. The diameter did not exhibit major impact on the gas yield. In contrast, gasification efficiency increased with increasing inner diameter in case of sawdust, while decreased in case of glucose. The cumulative degree of thermodynamic perfection was calculated on the bases of exergy and environmental impact. In this case, the thermodynamic perfection was higher, as the inner diameter of the reaction tube was increased from 3 mm to 9 mm [110]. In addition, Lu *et al.* [99] used SCW fluidized bed with a bed diameter of 30 mm and total length of 915 mm during SCWG of biomass and did not observed the plugging problem. So far, only a few studies have reported SCWG of biomass in fluidized bed, therefore, it is difficult to generalize the behavior of fluidized bed reactors.

The majority of SCWG studies have been carried out in small diameter and higher height reactors, therefore, the reactors exhibited very large surface to volume ratios. Generally, metals are used for reactor preparation and catalysts are used to carry out the SCWG of biomass, therefore, gasification exhibited major influence of the reactor materials. Antal *et al.* [111] suggested that nickel alloy tubes are not appropriate for use in SCWG of biomass due to their catalytic effects. Also, Yu *et al.* [112] conducted the experiment in a reactor with inner walls of Inconel 625 and suggested that reactor material promoted the water-gas shift reaction during SCWG. In contrast, Basu and Mettanant [19] reported that reactors made of 316 steel resulted in decreased catalytic activity, therefore, gasification efficacy was lower. The tannery waste treated with chromium had significantly decreased gas yield compared to no chromium treatment. Thus, one of the key issues for SCWG technology is that the reactor material should be compatible with catalysts as well as biomass, thus, needing further critical exploration in view of its commercial implications to increase the viability of SCWG technology.

4.4 Technical Challenges and Future Recommendations

SCWG is a demanding technology due to supercritical conditions, therefore, it requires advanced materials suitable for high pressure and temperature processes. Corrosion of the reactor material, char formation and plugging are key challenges in SCWG of biomass, which need to addressed prudently [7,21,51,113]. The precipitation of salts from organic feedstock during SCWG, combined with char causing plugging of the reactors [7,8,108,114-116]. The char formation and plugging issues can be overcome with rapid and homogenous heating using continuous reactor configurations [93,95,100,117,118]. Also, corrosion issues can be minimized using advanced materials for reactor design and pre-treatment of corrosive biomass [118]. Biomass feed in higher concentrations is another technical challenge and requires efficient pumping processes. Therefore, to overcome this issue, the feedstock has to be pumped in slurry phase [7,16,119]. To optimize SCWG process, use of efficient energy recovery equipment is inevitable. High efficiency of heat exchanger is required to heat water and mix it with the biomass for efficient hydrothermal degradation to gases.

SCWG is a promising and eco-friendly technology to produce wide range of biofuels and chemicals from wet biomass at critical point. The biomass with higher moisture content is a suitable reaction medium for SCWG. The biomass such as food and agricultural waste demands additional disposal cost, however, it can be directly used as feedstock in SCWG without any pre-treatment as compared to conventional gasification technology [120-122]. Moreover, the gas compression cost is also decreased due to H_2 production along with CO_2 separation from the SCWG products, which is an added advantage of high pressure SCWG. Thus, SCWG can be useful as a part of a bio-refinery approach under circular economy due to promising opportunities. For example, during SCWG process, syngas is the product, whereas water is a byproduct. Therefore, H_2 and CH_4 can be used for upgrading the transportation of fuels after CO_2 separation. In addition, the separated CO_2 can be sequenced during algae growth, while byproduct water can also be recycled for algae production and algae biomass can be used for different purposes [20,123-125].

4.5 Conclusions

SCWG is a promising technology for biofuel production using wet biomass feedstock. The industrialization of this process is gradually increasing and the process is compatible (or superior than) with other available technologies for H_2 production. The basic reaction pathways in the process have been well-known and the impact of components of biomass, i.e., cellulose, hemicellulose, lignin, proteins and lipids has been gradually understood. However, the biomass feeds are heterogeneous with a complex composition, therefore, several aspects have to be taken into account for effective SCWG process. SCWG trials of glucose and lignin concluded that H_2 yield decreases in the presence of phenol with glucose, while the char formation and plugging issues decrease with lignin. Also, during the SCWG of cellulose or glucose, H_2-rich gas mix can be obtained, whereas SCWG of lignin produces CH_4-rich gas mix. Studies on SCWG of biomass have also demonstrated the effect of salts and catalyzing hydrogen formation. Several studies have also been devoted to the optimization of operating conditions as well as scale up of the process.

This chapter has summarized the recent developments in the SCWG process of biomass. Several aspects for technology upgrading such as reactor configurations and operating conditions (temperature, pressure, feed concentration and catalyst), which have crucial role in maximizing H_2 production as well as gasification efficiency, have been discussed. The improvement in the knowledge is assessed in view of the hurdles to be overcome for a wide industrial application of SCWG technology.

References

1. McKendry, P. (2002) Energy production from biomass (part 1): overview of biomass. *Bioresource Technology*, **83**(1), 37-46.
2. Karthikeyan, O. P., Trably, E., Mehariya, S., Bernet, N., Wong, J. W. C., and Carrere, H. (2018) Pretreatment of food waste for methane and hydrogen recovery: A review. *Bioresource Technology*, **249**, 1025-1039.
3. Food Waste Harms Climate, Water, Land and Biodiversity - New FAO Report (2013) *Food and Agricultural Organization of the United Nations*. Online: http://www.fao. org/news/story/en/item/196220/icode/ (assessed 19th June 2018).
4. Mirabella, N., Castellani, V., and Sala, S. (2014) Current options for the valorization of food manufacturing waste: a review. *Journal of Cleaner Production*, **65**, 28-41.
5. Mehariya, S., Patel, A. K., Obulisamy, P. K., Punniyakotti, E., and Wong, J. W. C. (2018) Co-digestion of food waste and sewage sludge for methane production: Current status and perspective. *Bioresource Technology*, **265**, 519-531.
6. Kruse, A., Henningsen, T., Sınag, A., and Pfeiffer, J. (2003) Biomass gasification in supercritical water: Influence of the dry matter content and the formation of phenols. *Industrial & Engineering Chemis-*

try Research, **42**(16), 3711-3717.

7. Reddy, S. N., Nanda, S., Dalai, A. K., and Kozinski, J. A. (2014) Supercritical water gasification of biomass for hydrogen production. *International Journal of Hydrogen Energy*, **39**(13), 6912-6926.

8. Gasafi, E., Reinecke, M.-Y., Kruse, A., and Schebek, L. (2008) Economic analysis of sewage sludge gasification in supercritical water for hydrogen production. *Biomass and Bioenergy*, **32**(12), 1085-1096.

9. Brennan, L., and Owende, P. (2010) Biofuels from microalgae - A review of technologies for production, processing, and extractions of biofuels and co-products. *Renewable and Sustainable Energy Reviews*, **14**(2), 557-577.

10. Basu, P., and Mettanant, V. (2009) Biomass gasification in supercritical water - A review. *International Journal of Chemical Reactor Engineering*, **7**(1), DOI: https://doi.org/10.2202/1542-6580.1919.

11. Molino, A., Larocca, V., Chianese, S., and Musmarra, D. (2018) Biofuels production by biomass gasification: A review. *Energies*, **11**(4), 811.

12. Xu, X., Matsumura, Y., Stenberg, J., and Antal, M. J. (1996) Carbon-catalyzed gasification of organic feedstocks in supercritical water. *Industrial & Engineering Chemistry Research*, **35**(8), 2522-2530.

13. Leible, L., Kalber, S., Kappler, G., Lange, S., Nieke, E., Wintzer, D. F. B., and Furniss, B. (2006) Competitiveness and CO_2 mitigation costs of biogenic residues and waste for heat and power production. In: *Science in Thermal and Chemical Biomass Conversion*, Bridgwater, A. V., and Boocock, D. G. B. (eds.), CPL Press, UK.

14. Yen, H.-W., Yang, S.-C., Chen, C.-H., Jesisca, and Chang, J.-S. (2015) Supercritical fluid extraction of valuable compounds from microalgal biomass. *Bioresource Technology*, **184**, 291-296.

15. Yan, Q., Guo, L., and Lu, Y. (2006) Thermodynamic analysis of hydrogen production from biomass gasification in supercritical water. *Energy Conversion and Management*, **47**(11), 1515-1528.

16. Molino, A., Migliori, M., Blasi, A., Davoli, M., Marino, T., Chianese, S., Catizzone, E. and Giordano, G. (2017) Municipal waste leachate conversion via catalytic supercritical water gasification process. *Fuel*, **206**, 155-161.

17. Molino, A., Larocca, V., Valerio, V., Martino, M., Marino, T., Rimauro, J., and Casella, P. (2016) Biofuels and bio-based production via supercritical water gasification of peach scraps. *Energy & Fuels*, **30**(12), 10443-10447.

18. Molino, A., Larocca, V., Valerio, V., Rimauro, J., Marino, T., Casella, P., Cerbone, A., Arcieri, G. and Viola, E. (2018) Supercritical water gasification of lignin solution produced by steam explosion process on Arundo Donax after alkaline extraction. *Fuel*, **221**, 513-517.

19. Sanchez-Hernandez, A.M., Martin-Sanchez, N., Sanchez-Montero, M.J., Izquierdo, C., and Salvador, F. (2018) Effect of pressure on the gasification of dodecane with steam and supercritical water and consequences for H $_2$ production. *Journal of Materials Chemistry A*, **6**(4), 1671-1681.

20. Samiee-Zafarghandi, R., Karimi-Sabet, J., Abdoli, M. A., and Karbassi, A. (2018) Supercritical water gasification of microalga *Chlorella* PTCC 6010 for hydrogen production: Box-Behnken optimization and evaluating catalytic effect of MnO_2/SiO_2 and NiO/SiO_2. *Renewable Energy*, **126**, 189-201.

21. Nanda, S., Reddy, S. N., Dalai, A. K., and Kozinski, J. A. (2016) Subcritical and supercritical water gasification of lignocellulosic biomass impregnated with nickel nanocatalyst for hydrogen production. *International Journal of Hydrogen Energy*, **41**(9), 4907-4921.

22. Molino, A., Chianese, S., and Musmarra, D. (2016) Biomass gasification technology: The state of the art overview. *Journal of Energy Chemistry*, **25**(1), 10-25.

23. Correa, C. R., and Kruse, A. (2018) Supercritical water gasification of biomass for hydrogen production - Review. *The Journal of Supercritical Fluids*, **133**, 573-590.

24. Guo, Y., Wang, S. Z., Xu, D. H., Gong, Y. M., Ma, H. H., and Tang, X. Y. (2010) Review of catalytic supercritical water gasification for hydrogen production from biomass. *Renewable and Sustainable Energy Reviews*, **14**(1), 334-343.

25. Guo, L. J., Lu, Y. J., Zhang, X. M., Ji, C. M., Guan, Y., and Pei, A. X. (2007) Hydrogen production by biomass gasification in supercritical water: A systematic experimental and analytical study. *Catalysis Today*, **129**(3), 275-286.

26. Kruse, A., and Dahmen, N. (2015) Water - A magic solvent for biomass conversion. *The Journal of Supercritical Fluids*, **96**, 36-45.

27. Ni, M., Leung, D. Y. C., Leung, M. K. H., and Sumathy, K. (2006) An overview of hydrogen production from biomass. *Fuel Processing Technology*, **87**(5), 461-472.

28. Patel, S. K. S., Kumar, P., Mehariya, S., Purohit, H. J., Lee, J.-K., and Kalia, V. C. (2014) Enhancement in hydrogen production by co-cultures of *Bacillus* and *Enterobacter*. *International Journal of Hydrogen Energy*, **39**(27), 14663-14668.

29. Kumar, P., Chandrasekhar, K., Kumari, A., Sathiyamoorthi, E., and Kim, B. S. (2018) Electro-fermentation in aid of bioenergy and biopolymers. *Energies*, **11**(2), 343.

30. Rehan, M., Gardy, J., Demirbas, A., Rashid, U., Budzianowski, W. M., Pant, D. and Nizami, A. S. (2018) Waste to biodiesel: A preliminary assessment for Saudi Arabia. *Bioresource Technology*, **250**, 17-25.

31. Zhang, L., Champagne, P., and Xu, C. C. (2011) Supercritical water gasification of an aqueous by-product from biomass hydrothermal

liquefaction with novel Ru modified Ni catalysts. *Bioresource Technology*, **102**(17), 8279-8287.

32. Isikgor, F. H., and Becer, C. R. (2015) Lignocellulosic biomass: a sustainable platform for the production of bio-based chemicals and polymers. *Polymer Chemistry*, **6**(25), 4497-4559.

33. Bain, R. L., and Broer, K., 2011. Gasification. In: Brown, R. C., Stevens, C. (eds.), *Thermochemical Processing of Biomass*. John Wiley & Sons, United Kingdom, pp. 47-74.

34. Shen, Y., Jarboe, L., Brown, R., and Wen, Z. (2015) A thermochemical-biochemical hybrid processing of lignocellulosic biomass for producing fuels and chemicals. *Biotechnology Advances*, **33**(8), 1799-1813.

35. Naik, S., Goud, V. V, Rout, P. K., and Dalai, A. K. (2010) Supercritical CO_2 fractionation of bio-oil produced from wheat-hemlock biomass. *Bioresource Technology*, **101**(19), 7605-7613.

36. Zanzi, R., Sjostrom, K., and Bjornbom, E. (1993) Rapid pyrolysis of wood with application to gasification. In: *Advances in Thermochemical Biomass Conversion*, Bridgwater, A. V. (ed.), Springer, Netherlands, pp. 977-985.

37. Mitsuru, S., Tadafumi, A., and Kunio, A. (2004) Kinetics of cellulose conversion at 25 MPa in sub- and supercritical water. *AIChE Journal*, **50**(1), 192-202.

38. Kumar, P., Barrett, D. M., Delwiche, M. J., and Stroeve, P. (2009) Methods for pretreatment of lignocellulosic biomass for efficient hydrolysis and biofuel production. *Industrial & Engineering Chemistry Research*, **48**(8), 3713-3729.

39. Yoshida, Y., Dowaki, K., Matsumura, Y., Matsuhashi, R., Li, D., Ishitani, H., and Komiyama, H. (2003) Comprehensive comparison of efficiency and CO_2 emissions between biomass energy conversion technologies - position of supercritical water gasification in biomass technologies. *Biomass and Bioenergy*, **25**(3), 257-272.

40. Yoshida, T., and Matsumura, Y. (2001) Gasification of cellulose, xylan, and lignin mixtures in supercritical water. *Industrial & Engineering Chemistry Research*, **40**(23), 5469-5474.

41. Kruse, A., and Dinjus, E. (2007) Hot compressed water as reaction medium and reactant: Properties and synthesis reactions. *The Journal of Supercritical Fluids*, **39**(3), 362-380.

42. Klingler, D., and Vogel, H. (2010) Influence of process parameters on the hydrothermal decomposition and oxidation of glucose in sub- and supercritical water. *The Journal of Supercritical Fluids*, **55**(1), 259-270.

43. Goodwin, A. K., and Rorrer, G. L. (2008) Conversion of glucose to hydrogen-rich gas by supercritical water in a microchannel reactor. *Industrial & Engineering Chemistry Research*, **47**(12), 4106-4114.

44. Onda, A., Ochi, T., Kajiyoshi, K., and Yanagisawa, K. (2008) A new chemical process for catalytic conversion of d-glucose into lactic acid and gluconic acid. *Applied Catalysis A: General*, **343**(1), 49-54.

45. Srokol, Z., Bouche, A.-G., van Estrik, A., Strik, R. C. J., Maschmeyer, T., and Peters, J. A. (2004) Hydrothermal upgrading of biomass to biofuel; studies on some monosaccharide model compounds. *Carbohydrate Research*, **339**(10), 1717-1726.

46. Molino, A., Nanna, F., Villone, A., Iovane, P., Tarquini, P., Migliori, M., Giordano, G. and Braccio, G. (2014) Pressure and time effect over semi-continuous gasification of zootechnical sludge near critical condition of water for green chemicals production. *Fuel*, **136**, 172-176.

47. Hao, X., Guo, L., Zhang, X., and Guan, Y. (2005) Hydrogen production from catalytic gasification of cellulose in supercritical water. *Chemical Engineering Journal*, **110**(1), 57-65.

48. Azadi, P., Khan, S., Strobel, F., Azadi, F., and Farnood, R. (2012) Hydrogen production from cellulose, lignin, bark and model carbohydrates in supercritical water using nickel and ruthenium catalysts. *Applied Catalysis B: Environmental*, **117-118**, 330-338.

49. Azadi, P., and Farnood, R. (2011) Review of heterogeneous catalysts for sub- and supercritical water gasification of biomass and wastes. *International Journal of Hydrogen Energy*, **36**(16), 9529-9541.

50. Nanda, S., Isen, J., Dalai, A. K., and Kozinski, J. A. (2016) Gasification of fruit wastes and agro-food residues in supercritical water. *Energy Conversion and Management*, **110**, 296-306.

51. Nanda, S., Mohammad, J., Reddy, S. N., Kozinski, J. A., and Dalai, A. K. (2014) Pathways of lignocellulosic biomass conversion to renewable fuels. *Biomass Conversion and Biorefinery*, **4**(2), 157-191.

52. Kang, K., Azargohar, R., Dalai, A. K., and Wang, H. (2016) Hydrogen production from lignin, cellulose and waste biomass via supercritical water gasification: Catalyst activity and process optimization study. *Energy Conversion and Management*, **117**, 528-537.

53. Kang, S., Li, X., Fan, J., and Chang, J. (2013) Hydrothermal conversion of lignin: A review. *Renewable and Sustainable Energy Reviews*, **27**, 546-558.

54. Zhang, Y.-H. P. (2008) Reviving the carbohydrate economy via multi-product lignocellulose biorefineries. *Journal of Industrial Microbiology & Biotechnology*, **35**(5), 367-375.

55. Wahyudiono, T., K., M., S., and M., G. (2007) Decomposition of a lignin model compound under hydrothermal conditions. *Chemical Engineering & Technology*, **30**(8), 1113-1122.

56. Wahyudiono, Sasaki, M., and Goto, M. (2009) Conversion of biomass model compound under hydrothermal conditions using batch reactor. *Fuel*, **88**(9), 1656-1664.

57. Osada, M., Yamaguchi, A., Hiyoshi, N., Sato, O., and Shirai, M. (2012) Gasification of sugarcane bagasse over supported ruthenium catalysts in supercritical water. *Energy & Fuels*, **26**(6), 3179-3186.

58. Osada, M., Sato, T., Watanabe, M., Adschiri, T., and Arai, K. (2004) Low-temperature catalytic gasification of lignin and cellulose with a ruthenium catalyst in supercritical water. *Energy & Fuels*, **18**(2), 327-333.

59. Watanabe, M., Inomata, H., Osada, M., Sato, T., Adschiri, T., and Arai, K. (2003) Catalytic effects of NaOH and ZrO_2 for partial oxidative gasification of n-hexadecane and lignin in supercritical water. *Fuel*, **82**(5), 545-552.

60. Yanik, J., Ebale, S., Kruse, A., Saglam, M., and Yüksel, M. (2008) Biomass gasification in supercritical water: II. Effect of catalyst. *International Journal of Hydrogen Energy*, **33**(17), 4520-4526.

61. Yanik, J., Ebale, S., Kruse, A., Saglam, M., and Yuksel, M. (2007) Biomass gasification in supercritical water: Part 1. Effect of the nature of biomass. *Fuel*, **86**(15), 2410-2415.

62. Yong, T. L.-K., and Matsumura, Y. (2012) Reaction kinetics of the lignin conversion in supercritical water. *Industrial & Engineering Chemistry Research*, **51**(37), 11975-11988.

63. Zhang, L., Xu, C. C., and Champagne, P. (2010) Energy recovery from secondary pulp/paper-mill sludge and sewage sludge with supercritical water treatment. *Bioresource Technology*, **101**(8), 2713-2721.

64. Okuda, K., Umetsu, M., Takami, S., and Adschiri, T. (2004) Disassembly of lignin and chemical recovery - rapid depolymerization of lignin without char formation in water-phenol mixtures. *Fuel Processing Technology*, **85**(8), 803-813.

65. Fang, Z., Sato, T., Smith, R. L., Inomata, H., Arai, K., and Kozinski, J. A. (2008) Reaction chemistry and phase behavior of lignin in high-temperature and supercritical water. *Bioresource Technology*, **99**(9), 3424-3430.

66. Huelsman, C. M., and Savage, P. E. (2012) Intermediates and kinetics for phenol gasification in supercritical water. *Physical Chemistry Chemical Physics*, **14**(8), 2900-2910.

67. Huelsman, C. M., and Savage, P. E. (2013) Reaction pathways and kinetic modeling for phenol gasification in supercritical water. *The Journal of Supercritical Fluids*, **81**, 200-209.

68. Sato, N., Quitain, A. T., Kang, K., Daimon, H., and Fujie, K. (2004) Reaction kinetics of amino acid decomposition in high-temperature and high-pressure water. *Industrial & Engineering Chemistry Research*, **43**(13), 3217-3222.

69. Kruse, A., Krupka, A., Schwarzkopf, V., Gamard, C., and Henningsen, T. (2005) Influence of proteins on the hydrothermal gasification and liquefaction of biomass. 1. Comparison of different feedstocks.

Industrial & Engineering Chemistry Research, **44**(9), 3013-3020.

70. Castello, D., Kruse, A., and Fiori, L. (2015) Low temperature super-critical water gasification of biomass constituents: Glucose/phenol mixtures. *Biomass and Bioenergy*, **73**, 84-94.

71. Holliday, R. L., King, J. W., and List, G. R. (1997) Hydrolysis of vegetable oils in sub- and supercritical water. *Industrial & Engineering Chemistry Research*, **36**(3), 932-935.

72. Xu, D., Wang, S., Hu, X., Chen, C., Zhang, Q., and Gong, Y. (2009) Catalytic gasification of glycine and glycerol in supercritical water. *International Journal of Hydrogen Energy*, **34**(13), 5357-5364.

73. Antal, M. J., Allen, S. G., Schulman, D., Xu, X., and Divilio, R. J. (2000) Biomass gasification in supercritical water. *Industrial & Engineering Chemistry Research*, **39**(11), 4040-4053.

74. Diem, V., Boukis, N., Habicht, W., and Dinjus, E. (2003) The Catalytic Influence of the Reactor Material on the Reforming of Methanol in Supercritical Water. *6th International Symposium on Supercritical Fluids*, France. Online: http://www.isasf.net/fileadmin/files/Docs/Versailles/Papers/PRw5.pdf (assessed 20th June 2018).

75. Lu, Y. J., Guo, L. J., Ji, C. M., Zhang, X. M., Hao, X. H., and Yan, Q. H. (2006) Hydrogen production by biomass gasification in supercritical water: A parametric study. *International Journal of Hydrogen Energy*, **31**(7), 822-831.

76. Kruse, A., Meier, D., Rimbrecht, P., and Schacht, M. (2000) Gasification of pyrocatechol in supercritical water in the presence of potassium hydroxide. *Industrial & Engineering Chemistry Research*, **39**(12), 4842-4848.

77. Onwudili, J. A., and Williams, P. T. (2009) Role of sodium hydroxide in the production of hydrogen gas from the hydrothermal gasification of biomass. *International Journal of Hydrogen Energy*, **34**(14), 5645-5656.

78. Madenoglu, T. G., Boukis, N., Saglam, M., and Yuksel, M. (2011) Supercritical water gasification of real biomass feedstocks in continuous flow system. *International Journal of Hydrogen Energy*, **36**(22), 14408-14415.

79. Sheikhdavoodi, M. J., Almassi, M., Ebrahimi-Nik, M., Kruse, A., and Bahrami, H. (2015) Gasification of sugarcane bagasse in supercritical water; evaluation of alkali catalysts for maximum hydrogen production. *Journal of the Energy Institute*, **88**(4), 450-458.

80. Yoshida, T., Oshima, Y., and Matsumura, Y. (2004) Gasification of biomass model compounds and real biomass in supercritical water. *Biomass and Bioenergy*, **26**(1), 71-78.

81. Nguyen, H. T., Yoda, E., and Komiyama, M. (2014) Catalytic super-critical water gasification of proteinaceous biomass: Catalyst performances in gasification of ethanol fermentation stillage with batch and flow reactors. *Chemical Engineering Science*, **109**, 197-

203.

82. Furusawa, T., Sato, T., Sugito, H., Miura, Y., Ishiyama, Y., Sato, M., Itoh, N. and Suzuki, N. (2007) Hydrogen production from the gasification of lignin with nickel catalysts in supercritical water. *International Journal of Hydrogen Energy*, **32**(6), 699-704.

83. Yamaguchi, A., Hiyoshi, N., Sato, O., Bando, K. K., Osada, M., and Shirai, M. (2009) Hydrogen production from woody biomass over supported metal catalysts in supercritical water. *Catalysis Today*, **146**(1), 192-195.

84. Lu, Y., Li, S., Guo, L., and Zhang, X. (2010) Hydrogen production by biomass gasification in supercritical water over Ni/γAl$_2$O$_3$ and Ni/CeO$_2$-γAl$_2$O$_3$ catalysts. *International Journal of Hydrogen Energy*, **35**(13), 7161-7168.

85. Li, S., Guo, L., Zhu, C., and Lu, Y. (2013) Co-precipitated Ni-Mg-Al catalysts for hydrogen production by supercritical water gasification of glucose. *International Journal of Hydrogen Energy*, **38**(23), 9688-9700.

86. Onwudili, J. A., and Williams, P. T. (2013) Hydrogen and methane selectivity during alkaline supercritical water gasification of biomass with ruthenium-alumina catalyst. *Applied Catalysis B: Environmental*, **132-133**, 70-79.

87. Resende, F. L. P., and Savage, P. E. (2010) Effect of metals on supercritical water gasification of cellulose and lignin. *Industrial & Engineering Chemistry Research*, **49**(6), 2694-2700.

88. Schumacher, M., Yanık, J., Sınag, A., and Kruse, A. (2011) Hydrothermal conversion of seaweeds in a batch autoclave. *The Journal of Supercritical Fluids*, **58**(1), 131-135.

89. Youssef, E. A., Elbeshbishy, E., Hafez, H., Nakhla, G., and Charpentier, P. (2010) Sequential supercritical water gasification and partial oxidation of hog manure. *International Journal of Hydrogen Energy*, **35**(21), 11756-11767.

90. Lee, I.-G. (2011) Effect of metal addition to Ni/activated charcoal catalyst on gasification of glucose in supercritical water. *International Journal of Hydrogen Energy*, **36**(15), 8869-8877.

91. Lee, I.-G., and Ihm, S.-K. (2009) Catalytic gasification of glucose over Ni/activated charcoal in supercritical water. *Industrial & Engineering Chemistry Research*, **48**(3), 1435-1442.

92. Zhang, L., Champagne, P., and Xu, C. C. (2011) Screening of supported transition metal catalysts for hydrogen production from glucose via catalytic supercritical water gasification. *International Journal of Hydrogen Energy*, **36**(16), 9591-9601.

93. Susanti, R. F., Dianningrum, L. W., Yum, T., Kim, Y., Lee, B. G., and Kim, J. (2012) High-yield hydrogen production from glucose by supercritical water gasification without added catalyst. *International Journal of Hydrogen Energy*, **37**(16), 11677-11690.

94. Zhang, L., Xu, C. C., and Champagne, P. (2012) Activity and stability of a novel Ru modified Ni catalyst for hydrogen generation by supercritical water gasification of glucose. *Fuel*, **96**, 541-545.
95. Hendry, D., Venkitasamy, C., Wilkinson, N., and Jacoby, W. (2011) Exploration of the effect of process variables on the production of high-value fuel gas from glucose via supercritical water gasification. *Bioresource Technology*, **102**(3), 3480-3487.
96. Cao, C., Guo, L., Chen, Y., Guo, S., and Lu, Y. (2011) Hydrogen production from supercritical water gasification of alkaline wheat straw pulping black liquor in continuous flow system. *International Journal of Hydrogen Energy*, **36**(21), 13528-13535.
97. Madenoglu, T. G., Saglam, M., Yuksel, M., and Ballice, L. (2013) Simultaneous effect of temperature and pressure on catalytic hydrothermal gasification of glucose. *The Journal of Supercritical Fluids*, **73**, 151-160.
98. Lu, Y., Guo, L., Zhang, X., and Ji, C. (2012) Hydrogen production by supercritical water gasification of biomass: Explore the way to maximum hydrogen yield and high carbon gasification efficiency. *International Journal of Hydrogen Energy*, **37**(4), 3177-3185.
99. Lu, Y. J., Jin, H., Guo, L. J., Zhang, X. M., Cao, C. Q., and Guo, X. (2008) Hydrogen production by biomass gasification in supercritical water with a fluidized bed reactor. *International Journal of Hydrogen Energy*, **33**(21), 6066-6075.
100. Chen, Y., Guo, L., Cao, W., Jin, H., Guo, S., and Zhang, X. (2013) Hydrogen production by sewage sludge gasification in supercritical water with a fluidized bed reactor. *International Journal of Hydrogen Energy*, **38**(29), 12991-12999.
101. Peterson, A. A., Vontobel, P., Vogel, F., and Tester, J. W. (2008) In situ visualization of the performance of a supercritical-water salt separator using neutron radiography. *The Journal of Supercritical Fluids*, **43**(3), 490-499.
102. Matsumura, Y., Minowa, T., Potic, B., Kersten, S. R., Prins, W., van Swaaij, W. P., van de Beld, B., Elliott, D. C., Neuenschwander, G. G., Kruse, A. and Antal, Jr., M. J. (2005) Biomass gasification in near- and super-critical water: Status and prospects. *Biomass and Bioenergy*, **29**(4), 269-292.
103. D'Jesus, P., Boukis, N., Kraushaar-Czarnetzki, B., and Dinjus, E. (2006) Gasification of corn and clover grass in supercritical water. *Fuel*, **85**(7), 1032-1038.
104. Hao, X. H., Guo, L. J., Mao, X., Zhang, X. M., and Chen, X. J. (2003) Hydrogen production from glucose used as a model compound of biomass gasified in supercritical water. *International Journal of Hydrogen Energy*, **28**(1), 55-64.
105. Ji, C., Guo, L., Lu, Y., Pei, A., and Guo, X. (2007) Experimental investigation on hydrogen production by gasification of lignin in super-

critical water. *Taiyangneng Xuebao/Acta Energiae Solaris Sinica*, **28**(9), 961-966.
106. Byrd, A. J., Pant, K. K., and Gupta, R. B. (2008) Hydrogen production from glycerol by reforming in supercritical water over Ru/Al_2O_3 catalyst. *Fuel*, **87**(13), 2956-2960.
107. Hashaikeh, R., Butler, I. S., and Kozinski, J. A. (2006) Selective promotion of catalytic reactions during biomass gasification to hydrogen. *Energy & Fuels*, **20**(6), 2743-2747.
108. Kruse, A. (2008) Supercritical water gasification. *Biofuels, Bioproducts and Biorefining*, **2**(5), 415-437.
109. Matsumura, Y., and Minowa, T. (2004) Fundamental design of a continuous biomass gasification process using a supercritical water fluidized bed. *International Journal of Hydrogen Energy*, **29**(7), 701-707.
110. Lei, Y., Feng, X., and Min, S. (2007) Parameters optimization of hydrogen production from glucose gasified in supercritical water by equivalent cumulative exergy analysis. *Applied Thermal Engineering*, **27**(13), 2324-2331.
111. Antal, M. J., Mok, W. S. L., Roy, J. C., -Raissi, A. T., and Anderson, D. G. M. (1985) Pyrolytic sources of hydrocarbons from biomass. *Journal of Analytical and Applied Pyrolysis*, **8**, 291-303.
112. Yu, D., Aihara, M., and Antal, Jr., M. J. (1993) Hydrogen production by steam reforming glucose in supercritical water. *Energy & Fuels*, **7**(5), 574-577.
113. Nanda, S., Reddy, S. N., Hunter, H. N., Dalai, A. K., and Kozinski, J. A. (2015) Supercritical water gasification of fructose as a model compound for waste fruits and vegetables. *The Journal of Supercritical Fluids*, **104**, 112-121.
114. Akgul, G., and Kruse, A. (2012) Influence of salts on the subcritical water-gas shift reaction. *The Journal of Supercritical Fluids*, **66**, 207-214.
115. Reimer, J., Peng, G., Viereck, S., De Boni, E., Breinl, J., and Vogel, F. (2016) A novel salt separator for the supercritical water gasification of biomass. *The Journal of Supercritical Fluids*, **117**, 113-121.
116. Kruse, A., Forchheim, D., Gloede, M., Ottinger, F., and Zimmermann, J. (2010) Brines in supercritical biomass gasification: 1. Salt extraction by salts and the influence on glucose conversion. *The Journal of Supercritical Fluids*, **53**(1), 64-71.
117. Schubert, M., Müller, J. B., and Vogel, F. (2014) Continuous hydrothermal gasification of glycerol mixtures: Autothermal operation, simultaneous salt recovery, and the effect of K_3PO_4 on the catalytic gasification. *Industrial & Engineering Chemistry Research*, **53**(20), 8404-8415.
118. Vadillo, V., Sanchez-Oneto, J., Portela, J. R., and Martinez de la Ossa, E. J. (2013) Problems in supercritical water oxidation process and

proposed solutions. *Industrial & Engineering Chemistry Research*, **52**(23), 7617-7629.

119. Venkitasamy, C., Hendry, D., Wilkinson, N., Fernando, L., and Jacoby, W. A. (2011) Investigation of thermochemical conversion of biomass in supercritical water using a batch reactor. *Fuel*, **90**(8), 2662-2670.

120. Albarelli, J. Q., Mian, A., Santos, D. T., Ensinas, A. V, Marechal, F., and Meireles, M. A. A. (2015) Valorization of sugarcane biorefinery residues using supercritical water gasification: A case study and perspectives. *The Journal of Supercritical Fluids*, **96**, 133-143.

121. Wan, W. (2016) An innovative system by integrating the gasification unit with the supercritical water unit to produce clean syngas for solid oxide fuel cell (SOFC): System performance assessment. *International Journal of Hydrogen Energy*, **41**(48), 22698-22710.

122. Gumisiriza, R., Hawumba, J. F., Okure, M., and Hensel, O. (2017) Biomass waste-to-energy valorisation technologies: a review case for banana processing in Uganda. *Biotechnology for Biofuels*, **10**(1), 11.

123. Miguel, C. R., Calado, S. P., Alberto, R., Lopes, D. S. T., Hugo, M. V., and Yolanda, S. (2010) Supercritical fluid extraction of lipids from the heterotrophic microalga *Crypthecodinium cohnii. Engineering in Life Sciences*, **10**(2), 158-164.

124. Molino, A., Marino, T., Larocca, V., Casella, P., Rimauro, J., Cerbone, A., and Migliori, M. (2018) Supercritical water gasification of *Scenedesmus Dimorphus* μ-algae. *International Journal of Chemical Reactor Engineering*, **15**(4), DOI: https://doi.org/10.1515/ijcre-2016-0218.

125. Lopez Barreiro, D., Bauer, M., Hornung, U., Posten, C., Kruse, A., and Prins, W. (2015) Cultivation of microalgae with recovered nutrients after hydrothermal liquefaction. *Algal Research*, **9**, 99-106.

Chapter 5

Production of Biodiesel from Green Microalgae

Samuel Kofi Tulashie
University of Cape Coast, College of Agriculture and Natural Sciences, School of Physical Sciences, Department of Chemistry, Industrial Chemistry Section, Cape Coast, Ghana
stulashie@ucc.edu.gh

5.1 Introduction

The key part of all energy consumed worldwide comes from fossil sources (petroleum, coal and natural gas). However, these sources are limited, and will be exhausted in the near future. Thus, looking for alternative sources of new and renewable energy such as hydro, biofuel, wind, solar, geothermal, hydrogen and nuclear is of vital importance. Continued and increasing use of petroleum will intensify local air pollution and magnify the global warming problems caused by CO_2 [1].

Alternative new and renewable fuels have the potential to solve many of the current social problems and concerns, from air pollution and global warming to other environmental improvements and sustainability issues. There is a need to switch from conventional fuel sources (fossil fuels) to alternative fuel sources [2,3]. Biofuel, an alternative biological fuel, is made from renewable biological sources such as plants. It is biodegradable and nontoxic, has low emission profiles and is environmentally beneficial [4]. One of the sources of biofuel which has gained utmost attention is green microalgae for biodiesel production. This is due to several advantages portrayed by green microalgae. Some of these advantages are high growth rate [2], short time to maturity stage [5], large biomass production, lesser land area requirement [6-8] and biodegradability of the produced biodiesel [9]. Other major advantages include the arrangement of fatty acid containing compounds which are the best component for the production of biodiesel, ability to control them by varying the growth conditions [10-14], ability to generate numerous diverse categories of lipids based on their species [3], their

Biofuels, edited by Vikas Mittal
© 2018 Central West Publishing, Australia

potential to fix approximately 40-50% of the overall global organic carbon, such as CO_2, via photosynthesis, though they constitute only 0.2% of global biomass [15]. In addition to the high photosynthetic efficiency of microalgae, mass cultivation of microalgae is believed to efficiently reduce the carbon dioxide emission to the atmosphere, thus, reducing the impact of global warming [16,17]. Green microalgae also releases significant amount of oxygen to the atmosphere, thereby supporting the bulk of life on the planet, as stated below:

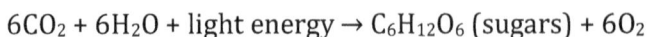

$$6CO_2 + 6H_2O + \text{light energy} \rightarrow C_6H_{12}O_6 \text{ (sugars)} + 6O_2$$

Based on this knowledge, research into biomass (green microalgae) exploitation for biofuel is conducted. The interest in microalgae for oil production is due to the high lipid content of some species, and to the fact that lipid synthesis, especially of the non-polar triacylglycerols (TAGs), which are the best substrate to produce biodiesel, can be regulated by varying growth conditions. The potential production, optimization and characterization of oil from microalgae have been reported in several studies [18-26]. For instance, Chisti reported the oil contents in fourteen different microalgae species to be at 15-75% dry weight [12]. Moreover, oil produced from microalgae was indicated to be 15-300 times greater than other conventional crops on the basis of area [27]. Conversely, the use of different solvents and different combinations of solvents for the algae oil extraction has also been employed in several studies. Folch *et al.* [28] method (2:1 v/v chloroform-methanol), Bligh and Dyer [29] method (1:2 v/v chloroform- methanol) and Matyash *et al.* method which is the modification of Folch as well as Bligh and Dyer methods with methyl-tert-butyl ether represent typical samples of such blends [18]. However, these combinations or methods have been mentioned to be environmentally unfavourable when used on a large scale [30]. As a consequence, high-class solvents such as ethanol, butanol, hexane, esters, ethers and their corresponding mixtures have been investigated [31]. For instance, 7.3% and 9.2% of alga oils were generated from two microalgae species, namely, *spirogyra sp.* and *oedogonium sp.* respectively, using hexane-ether solution [32]. Besides, 62.04±2.42%, and 40.71%±4.26% algae oils were extracted from the microalga *scenedesmus obliquus* using ethanol and hexane respectively [33]. Although ethanol resulted in higher yield, however, it also extracted other green pigments, which required additional purification, thus, making the use of hexane more suitable. The usage of hexane alone has also been reported to be feasible and effective [34,35]. Furthermore, hexane is cheaper, therefore, can be used in a large scale or on commercial scale to re-duce production cost.

This work investigated the production of biodiesel from green microalgae applying hexane as an extracting solvent via Soxhlet ex-tractor. Inorganic fertilizer was employed as nutrient treatment during culturing of the freshwater green microalgae, which were characterized using a digital microscope. The growth rates and algae oil yields are presented and discussed. The first part of this work demonstrates the production of biodiesel from green microalgae as a model case, as the cultivation is conducted at the laboratory scale (batch reactor regime case study). Afterwards, the second section of this work comprises of a comprehensive upscaling/pilot plant (commercialization case) approach. Various challenges arising out of the prototype system as well as their solutions, which mostly in-volve large-scale production regimes, are exploited and examined.

5.2 Materials and Methods

5.2.1 Materials and Equipment

Hexane, ethanol (70%), iodine tincture and diethyl ether were ac-quired from the chemical storeroom of the Chemistry Department of the University of Cape Coast, Ghana. Yara vita fertilizer was pur-chased from a licensed fertilizer retailer in Cape Coast, Kotokruba market. Whatman filter paper number 4, pipette, hot plate, beakers, microscope (having > 1000x magnification), test tubes, two Soxhlet extractor set ups, Gilson pipette, 500 ml flat bottom flasks, centri-fuge (brand 5702R), zooplankton and phytoplankton net, rotary evaporator, Eppendorf tubes and haemocytometer were procured from Chemistry and Microbiology departments of the University of Cape Coast.

5.2.2 Sampling and Sample Preparation

The sample was collected from the science botanical garden pond in the University of Cape Coast, Ghana. The nature and location of the pond demonstrated the presence of phytoplankton (freshwater green microalgae). The sampling process was conducted two times. The first sample was collected for centrifugation and serial dilution for the growth process. In this case, 1 L of pond water, which contained the microalgae, was taken to the laboratory. The algae were concentrated and prepared as stock for the serial dilution by centrifuging at 1000 rpm for 4 minutes [26]. The second sample (the raw sample) was collected for sun drying without centrifugation. For this purpose, 5 L of pond water was collected and sun dried.

Isolation of Algal Species Using Serial Dilution Method

To achieve minute and fine molar concentrations of chemicals and compositions, the chemists in chemical laboratories have often applied serial dilution method. This study also applied the same method to isolate (separate) and categorize individual colonies present in the collected sample. Isolation and collection of a particular colony from the freshwater green microalgae was cost-effective rather than buying these samples.

To start the serial dilution process, all apparatuses used for the process were first carefully cleaned to avoid contamination. Ten (10) test tubes were organized in a test tube rack and filled with 9 mL of distilled water. 1 mL of the centrifuged stock sample, which had higher concentration of microalgae, was conveyed into the first test tube. Then 1 mL was taken from the first test tube and poured into the second test tube. Similarly, 1 mL from the second test tube was taken and poured into the third test tube. The process was repeated for the ten test tubes as depicted in Figure 5.1. Once the process was completed, three test tubes were randomly selected to serve as inoculum for the culturing process. The procedure was repeated five times for four different nutrient concentration treatments and a control setup.

Microalgae were cultivated over 7 successive days, in 500 mL batch bioreactor illuminated from above with a 5200-lux "daylight" and/or fluorescent light. Reproducibility experiments were conducted by using three replicates for each nutrient treatment and kept under constant aeration to guarantee reproducibility. The culture growth was monitored closely to determine the variation of the number of cells throughout the growth period. Haemocytometer was used for counting of the algae cells. A growth curve was derived for the respective treatment and regulated to a model to differentiate between the different stages of growth. The growth rate was evaluated by applying the exponential phase of the growth curve, which characterizes the number of cellular splitting per day and subsequent replication time. Also, the stipulated division time or generation time was determined from the growth rate as reported by Stein [36].

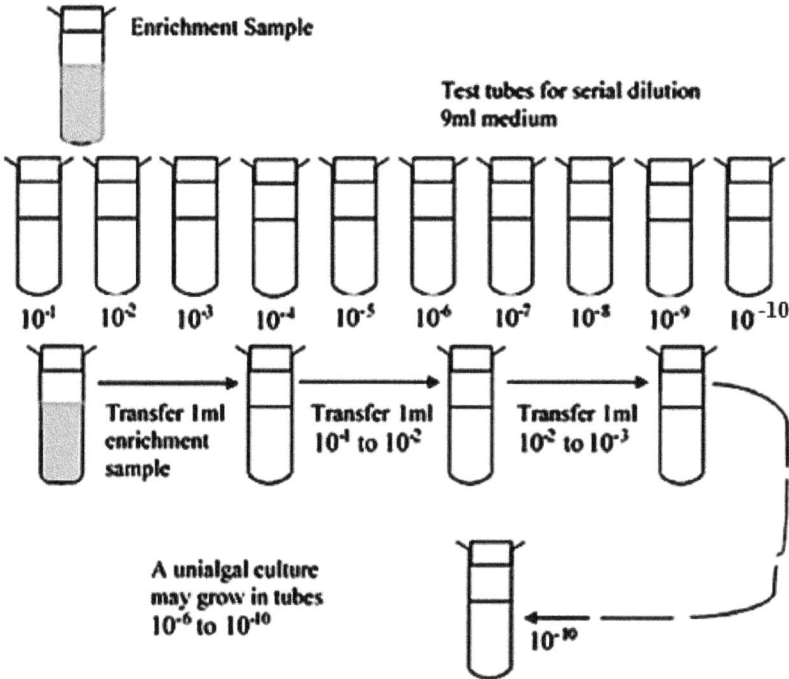

Figure 5.1 Schematic representation of the serial dilution method.

Sterilization

Sterilization process is conducted before carrying out the operations such as media preparation, batch cultivation as well as continuous stirred tank bioreactor (CSTR) cultivation, which is used for commercial production of green microalgae. Sterilization process eradicates, removes, destroys and deactivates all microorganisms. In this process, heat is applied in the form of steam, thus, it is also known as thermal destruction. The rate of elimination of the microorganism by heat can be described in terms of first order kinetics. The elimination rate is almost proportional to the number of surviving cells present in the system being heated. The rate of destruction may be expressed as:

$$\frac{dN}{dt} = -kN \tag{1}$$

where N is the number of viable organisms at any time, t is time in minutes and k is the rate constant. The rate constant is dependent on the type of microorganism and the ambient temperature. Thus, the equation 1 can be modified as

$$\frac{dN}{N} = -kdt \tag{2}$$

$$\int_{N_i}^{N} \frac{dN}{N} = -k \int_{t=0}^{t} dt \tag{3}$$

$$In\ [N]_{N_i}^{N} = -k \int_{t=0}^{t} dt \tag{4}$$

$$In\ [N]_{N_i}^{N} = -kt \tag{5}$$

$$In\ N - In\ N_i = -kt \tag{6}$$

$$\frac{N}{N_i} = e^{-kt} \tag{7}$$

$$-In\left(\frac{N}{N_i}\right) = kt \tag{8}$$

$$In\left(\frac{N_i}{N}\right) = kt \tag{9}$$

Microbiologists normally refer to decimal reduction time, D, which is the time in minutes during which the initial number of microorganisms is reduced to $1/10^{th}$. This represents 90% reduction. Replacing for the ratio of (N/N_i) in the aforementioned equation, we obtain:

$$\frac{N}{N_i} = \frac{1}{10} = e^{-kD} \tag{10}$$

$$In\left(\frac{1}{10}\right) = -kD \tag{11}$$

$$-\left[In\left(\frac{1}{10}\right)\right] = kD \tag{12}$$

$$In\ 10 = kD \tag{13}$$

$$D = \frac{2.303}{k} \tag{14}$$

From equation 9,

$$t = \frac{In \ (N_i/N)}{k} \tag{15}$$

However, $In = 2.303 \ log_{10}$

$$t = \frac{2.303 \ log_{10} \ (N_i/N)}{k} \tag{16}$$

Substituting equation 14 into 16 gives

$$t = D \ log_{10} \left(\frac{N_i}{N}\right) \tag{17}$$

A graphical plot of log (N_i/N) versus t gives a straight line. Herein, the D-value signifies the time in minutes required to decrease the initial spore population by 10^1, i.e. one log cycle. Substituting into equation 17 and solving for t gives

$$t = \frac{In \ 10}{k} \tag{18}$$

For complete elimination of spore or microorganisms population represents 10^{12} for the case of sterilization, which implies

$$T = \frac{In \ 10^{12}}{k} \tag{19}$$

$$T = \frac{12 \times 2.303}{k} = 12 \ D \tag{20}$$

Herein, equation 20 gives *12D* and which is regularly called *12D* concept, which is a mathematical concept found experimentally to provide effective sterilization. The thermal destruction procedure time required for sterilization is usually based on the necessity to eliminate all the spores/microorganisms of *Clostridium botulinum*, which is used as a worst case scenario, where the poisonous neuro-toxin can easily multiply into a large population. As a safety meas-ure, a very large initial population of the organism is assumed to be present and the heat treatment process should reduce the number of initial spores by a factor of 10^{12}.

Moreover, to be able to calculate for thermal process time and adequacy of heat treatment, the following equation can be applied

$$F_0 = t_1 \cdot 10^{(T_1 - 250)/Z} + t_2 \cdot 10^{(T_2 - 250)/Z} + t_3 \cdot 10^{(T_3 - 250)/Z} \qquad (21)$$

For illustration, if a given reactor is subjected to steam for sterilization with the temperature variation for different periods from 0-20 min at 160 °F (t_1 = 20 min and T_1 = 160 °F), 20-40 min at 210 °F (t_2 = 20 min and T_2 = 210 °F) and 40-73 min at 230 °F (t_3 = 33 min and T_3 = 230 °F), the total F_0 value of the heat treatment may be calculated by taking F_0 for *Clostridium botulinum* as 2.50 minutes representing the worst case scenario for contamination and 18 *F* as z value representing the number of degrees required for a specific thermal elimination time curve to pass through one log cycle. Substituting temperatures and corresponding time in the above equation, we obtain:

$$F_0 = 20 \cdot 10^{(160 - 250)/18} + 20 \cdot 10^{(210 - 250)/18} + 33 \cdot 10^{(230 - 250)/18}$$
$$(22)$$

$$F_0 = 2.68 \ minutes \qquad (23)$$

where F_o value symbolizes the number of minutes to destroy a stipulated number of an initial population of an organism with specific z value at a specific temperature.

As the required F_0 value is 2.50 minutes and calculated F_0 value is 2.68 minutes, thus, the steam treatment provided by the different time-temperature combinations is considered to be adequate.

Media Preparation and Batch Culture

Inorganic nutrients were applied in this study. Yara vita fertilizer of the following composition was used to prepare nutrients of known concentrations: 5% boron (B), 5% zinc (Zn), 0.1% copper (Cu), 0.1% manganese (Mn), 0.1% molybdenum (Mo), 5% nitrogen (N), 7.5% phosphate (P_2O_5), 5% potash (K_2O), 5% magnesium (Mg) and 5% sulphur (S). This served as a media for the cells of the freshwater microalgae to use as source of food for growth/multiplication. Any essential alterations (addition of fresh water/evaporation) completed would also lead to nutrient concentrations changes in the system.

For this purpose, 0.175 g, 0.35 g, 2 g and 4 g of fertilizer were weighed using weighing balance and each sample was dissolved in a

200 mL beaker. Small amount of distilled water was added and the content was stirred to dissolve the powdery fertilizer. After the dissolution process, more distilled water was added to top up the beaker to the 200 mL mark. The volume of the mixture was then suction filtered to remove the particles remaining in water after dissolution. The 200 mL nutrient solution was poured into a 500 mL batch bioreactor. 300 mL of distilled water was introduced to make up 500 mL and was repeated for two more batch bioreactors. The content of the three test tubes of the serial dilution process were subsequently charged to the batch bioreactor.

Figure 5.2 shows the illustration of a batch bioreactor, which was used for the batch culture of the green microalgae. In the figure, X

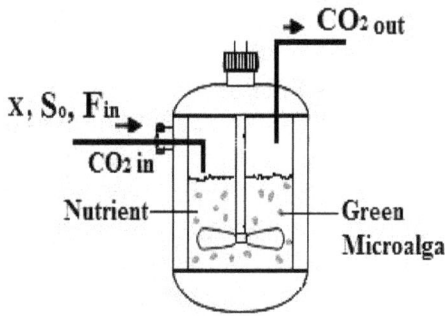

Figure 5.2 A batch bioreactor.

depicts biomass input, S_o represents substrate or nutrient, F_{in} represents flow rate, CO_2 in is the carbon dioxide input stream and CO_2 out is the carbon dioxide output stream. An ideal perfectly mixed batch bioreactor is largely homogeneous as a consequence of intensive agitation. The material balance for batch cultivation is given by the rate of accumulation of the product, which is equal to the rate of formation of the product due to chemical reaction. The most general form of this equation for mass balance is

$$\frac{d(V_R C)}{dt} = V_R r \tag{24}$$

$$\frac{V_R dc}{dt} + \frac{c dV_R}{dt} = V_R r \tag{25}$$

$$\frac{V_R dc}{dt} + \frac{c dV_R}{dt} = V_R r \tag{26}$$

$$\frac{V_R dc}{dt} = V_R r - \frac{c dV_R}{dt} \tag{27}$$

For a reactor without a recycle system, $\frac{c dV_R}{dt} = 0$

$$\frac{dc}{dt} = r \tag{28}$$

$$\frac{dX}{dt} = \mu x \tag{29}$$

where μ is the specific growth rate, c is the amount of the component, V_R is the total volume of the culture in the bioreactor and r is the reaction rate.

The nutrient preparation was repeated for the four masses and a control setup was included without addition of nutrients. The concentrations of the measured masses of the fertilizer were calculated to be 350 mg/L, 700 mg/L, 4000 mg/L, 8000 mg/L and 0 mg/L for the 0.175 g, 035 g, 2 g, 4 g and control, respectively. In the process of preparing a stock solution, which contained a mixture of compounds as in the case of Yara vita fertilizer, it was prudent to dissolve each mass in a minimal volume of water before mixing, subsequently combine and dilute to volume. After the nutrient and the samples were combined in the batch bioreactor, the treatment was monitored every 24 h for cell count using a microscope and a haemocytometer. All equipment used in the growth process of the green microalgae were sterilized.

For future batch cultivation of the green microalgae, it is recommended to use a modified batch bioreactor due to several challenges encountered in the process. Some of the challenges confronted in the batch cultivation process include foam formation in the reactor, absence of sterilization piping, absence for a liquid level indicator for filling and no indication for the determination of reactor weight. Therefore, with all the aforementioned challenges, it is better to employ a modified batch bioreactor, which is depicted in Figure 5.3. Operation related to bioreactors includes gas/liquid contacting, online sensing of concentrations, mixing, heat transfer, foam control and feed of nutrient/substrate such as those for pH regulator. In the modified batch bioreactor, contact cooling coils are placed between the baffle and tank wall or connect to the top to reduce opening below the water level. Water level and bottom-entering mixers are used frequently in these bioreactors. Cooling coils must be used for larger tanks, as the heat transfer area of a jacket is insufficient for cooling from sterilization temperature to working temperature in a rational period of time. Some of the features applied in the modified batch bioreactor are (i) a bypass valve in the air system, which allows diversion of air so as to minimize foaming, since the effective volume of the liquid in the bioreactor would increase and subsequently affect the hydrodynamic effect in the bioreactor and eventually mass transfer, (ii) all piping sterilized by the use of steam and protected by steam until put into use, (iii) the level of liquid when filling the vessel determined by reference to a calibration chart based on points in the tank and (iv) the weight of tank contents evaluated by the hydrostatic balance against air bubble slowly through the sparger.

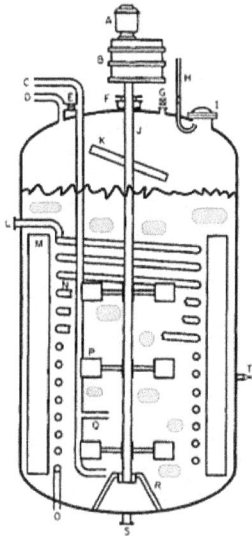

Figure 5.3 Modified batch bioreactor: A = agitator motor; B = speed reduction unit; C = air inlet; D = air outlet; air bypass valve; F = shaft seal; G = sight glass with light; H = sight glass clean-off line; I – man hole with sight glass; J = agitator shaft; K = paddle to break foam; L = cooling water outlet; M = baffle; N =cooling coils; O = cooling water inlet; P = mixer; Q = sparger; R = shaft bearing and bracket; S = outlet; T = sample valve [37].

Commercialization of Green Microalgae Cultivation

Algae are basically a large and varied group of simple and normally autotrophic organisms, ranging from unicellular to multicellular forms. Therefore, algae have the potential to produce significantly large amount of biomass and lipids per hectare than any terrestrial biomass. Also, these can be cultivated on minimal land, therefore, would not compete with foods or other crops.

For commercial cultivation trial of green microalgae, the continuous stirred tank bioreactor (CSTR) was employed for laboratory pilot-plant. The configuration of a typical well-mixed CSTR is shown in Figure 5.4. In the depicted CSTR systems, X indicated algae, S is the substrate/nutrient and F is the flow rate of the nutrient/substrate concentration into the bioreactor.

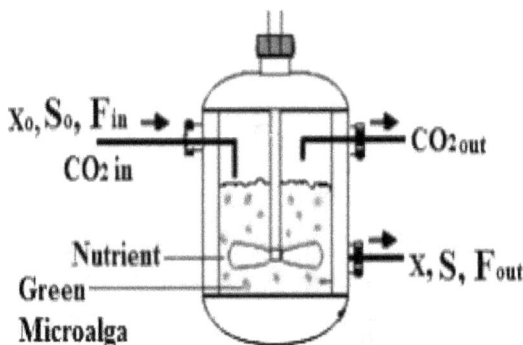

Figure 5.4 A schematic of a continuous stirred tank bioreactor.

The material balance for the continuous cultivation can be written as

$$\frac{d}{dt}(V_R c) = F_o c_o - F_i c_i - V_R r \tag{30}$$

In order to maintain the volume within the vessel, the volumetric flow rates, F_{in} and F_{out} into and from the vessel are maintained constant by

$$F_o = F_i = F \tag{31}$$

$$\frac{d}{dt}(V_R c) = F(c_o - c_i) - V_R r \tag{32}$$

For a reactor without a recycle system, $\frac{c dV_R}{dt} = 0$

$$\frac{dc}{dt} = \frac{F}{V_R}(c_o - c_i) - r \tag{33}$$

However, $F/V_R = D$ = dilution rate

$$\frac{dc}{dt} = D(c_o - c_i) - r \tag{34}$$

For biomass

$$\frac{dX}{dt} = D(X_o - X_i) - \mu X \tag{35}$$

There are several viable and practical commercial ways of algae cultivation. The comparison of the properties of these different large-scale algal cultivation methods is summarized in Table 5.1.

Table 5.1 Comparison of the properties of different large-scale algal cultivation methods (adapted from Reference 38)

Reactor type	Light usage efficiency	Temperature control	Mixing	Hydrodynamic stress on algae	Sterility	Scaleup
Stirred tank reactor	Fairly-good	Excellent	Largely uniform	High	Easily attainable	Difficult
Circular stirred ponds	Fairly-good	None	Fair	Low	None	Very difficult
Paddle-wheel Raceway ponds	Fairly-good	None	Fairly-good	Low	None	Very difficult
Tanks	Very poor	None	Poor	Very low	None	Very difficult

Preparation of the Sample for Haemocytometer Count

1 mL of medium was drawn from the batch bioreactor with the help of a pipette containing the cells of the freshwater microalgae and was appropriately prepared before applying onto the haemocytometer. Haemocytometer was cleaned using 70% ethanol. The shoulders of the haemocytometer were moisturized and the coverslip was affixed using mild pressure and tiny circular motions. The phenomenon of Newton's rings was observed when the coverslip was rightly affixed, hence, the depth of the chamber was confirmed.

Preparation of Cell Suspension

It was ensured that the cell suspension to be counted was properly mixed by tenderly mixing the flask holding the cells. 1 mL of the cell suspension was sampled out and poured in an Eppendorf tube before the cells could settle. Pipetted cells were mixed again tenderly to avoid lysing of the cells. Approximately 100 µL was drawn and placed into a new Eppendorf and subsequently two (2) drops of iodine tincture were added to the Eppendorf tube. The contents were mixed gently again to aid picture the cells in the counting procedure. Gilson pipette was used to sample the cell suspension containing iodine tincture. The haemocytometer was carefully filled by tenderly placing the end of the Gilson tip at the brink of the chambers. Maximum care was taken in the process not to exceed the capacity of the chamber.

The sample was taken out from the pipette by the application of the capillary action and care was taken for the sample to flow to the brinks of the grooves only. The pipette was re-loaded and filled for the second chamber. The haemocytometer was placed on the microscope hand. The microscope was focused on the grid lines of the haemocytometer employing the 10x objective lens. The set of 16-corner square of the haemocytometer was focused on, as shown by the circle in Figure 5.5.

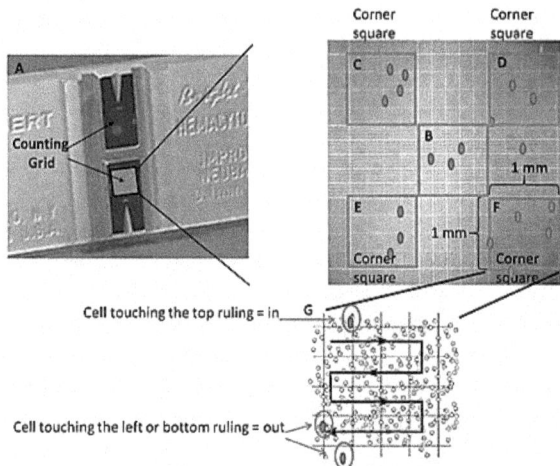

Figure 5.5 Depiction of the haemocytometer applied in counting of the algal cells [39].

Calculation of Cell Density

To evaluate cell density, certain measurements were required. These included: number of cells counted in a square, area of the square, height of the sample and dilution factor. The objective was to find the number of cells in 1 mL of original solution. The algal cells were seen as blackish spot and occasionally observed in colony. The cell concentrations were calculated as expressed in equation (36) as

$$\frac{cell\ counts\ x\ dilution\ factor}{Area\ (haemocytometer)x\ depth\ (haemocytometer)}\left(\frac{cells}{ml}\right) \tag{36}$$

The cell count was recorded every 24 h for 7 days. The data resulted from the cell count was subsequently used to obtain a growth curve for the freshwater microalgae.

5.2.3 Identification of Algae

The cultured freshwater green microalgae were visually observed and characterized using a microscope. The identification was performed with the aid of a phytoplankton identification manual [40]. Under the microscope, most of the cells were observed as green-like algae (Figure 5.6), which enabled their categorization under the chlorophyte (green microalgae) type.

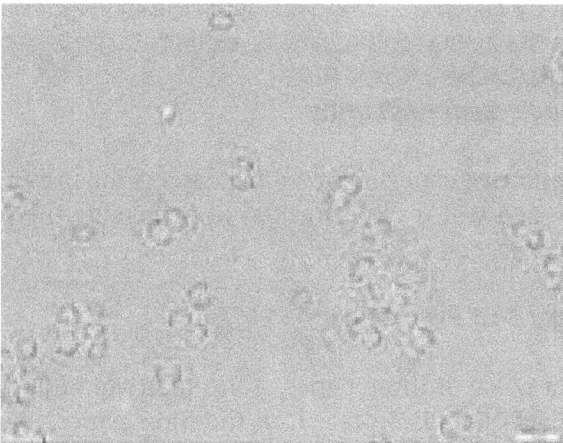

Figure 5.6 Slide showing an image of green microalgae as observed under a microscope.

Dewatering

Dewatering removes/separates water from the freshwater microalgae harvested from a freshwater pond. This was conducted in two ways, firstly, dewatering process where the microalgae were extracted from the medium using a centrifuge (5702 R) to concentrate the biomass and was eventually sun dried to remove the remaining moisture from the microalgae in a metallic tray. Secondly, after the freshwater microalgae was sampled from the freshwater pond with the aid of the zooplankton and phytoplankton, 5 L of the microalgae was sun dried for five days at ambient temperature between 25 to 28 °C without centrifugation in three metallic trays.

5.2.4 Lipid Extraction Using Hexane as Solvent

The sun-dried algae were crushed into a fine powder. The algae powder was subsequently dispensed into a paper container and enclosed to withstand any solid algae discharge. The container was arranged within an extraction chamber and successfully prepared for hexane extraction, as depicted in Figure 5.7. Two Soxhlet extraction set-ups were prepared for the extraction process. 150 mL of hexane was added to the distillation flask of the first set-up, which contained 15 g of the centrifuged cultured dry sample, whereas 300 mL of hexane was added to the second set-up, which contained 30 g of the raw dry sample. Extraction was carried out at 65 °C for 60 cycles in 3 h. The crude algal oil was collected by evaporative loss of hexane from the hexane/oil reservoir using rotary evaporator [41].

Tests for the Presence of Lipids

This is a qualitative experiment, which depicts the presence of oil in a substance. The test was performed to determine the presence of lipids in the extracted algae oil.

Translucency Test Procedure

It involved a piece of filter paper, a hot plate and ether solvent. A drop of the algae oil was placed on the filter paper. Subsequently, the filter paper was placed on the hot plate and heated to 60 °C for 5 minutes. The filter paper was removed and immersed in ether, after which it was air dried and the spot examined for translucency.

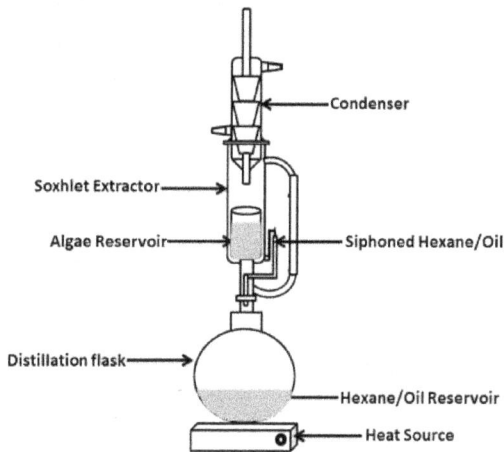

Figure 5.7 Soxhlet extraction set up for the algae oil extraction process.

5.3 Results and Discussion

5.3.1 Batch Reactor Case Study

The first part of this section involves the batch culture of the green microalgae and the results obtained for lipid extraction processes. The second section discusses the constraints associated with the scaleup or commercialization of algal biofuel.

Serial dilution method resulted in the growth of unialgal species of freshwater microalgae in the batch reactor. The cells were cultivated in a period of 24-264 h at an increase of 24 h. The isolation process applying the serial dilution method was conducted to realize a single or two species of freshwater microalgae growing in at least one of the batch reactors. Table 5.2 depicts the growth rates of cultured algae treated with variable amounts of Yara vita nutrients (0 g, 0.175 g, 0.35 g, 2 g and 4g) labelled as A, B, C, D and E respectively. It was observed that as the nutrients concentration increased, the growth rates normally increased throughout a specific period from left to right of the table. The algae growth rates for the control, 0.175 g and 0.35 g were comparable with differences of ±0.000 to ±0.009 (0%-22.5%). For instance, the algae growth rates for the control, 0.175 g and 0.35 g at 24 h were 0.040 cell/h, 0.031 cell/h and 0.031 cell/h respectively, while at 240 h were 0.003 cell/h, 0.003 cell/h and 0.004 cell/h respectively. Conversely, the nutrient concentrations 2 g and 4 g resulted in exponential algae growths as likened to the control, 0.175 g, and 0.35 g in the entire durations. This outcome is depicted in Figure 5.8, which suggests the growth phases for the cultured microalgae species (lag, exponential, phase of declining and death phase).

Table 5.2 The growth rates of cultured algae treated with variable amounts of Yara vita nutrients [39]

Duration culture/h	Control (A)	Growth rate (cell/h)	0.175 g (B)	Growth rate (cell/h)	0.35g (C)	Growth rate (cell/h)
24	-	-	-	-	-	-
48	-	-	-	-	-	-
72	2.9±1.2	0.040	2.2±0.8	0.031	2.2±1.0	0.031
96	1.5±0.9	0.015	1.5±0.4	0.016	1.7±0.8	0.018
120	1.1±0.3	0.009	1.7±0.4	0.014	1.6±0.1	0.013
144	1.4±0.2	0.010	2.9±	0.020	1.6±1.2	0.011
168	0.9±0.4	0.005	1.0±0.2	0.006	1.3±0.6	0.008
240	0.7±0.1	0.003	0.8±0.1	0.003	0.7±0.2	0.004
264	0.7±0.1	0.003	1.6±1.4	0.006	1.0±0.1	0.004

Duration culture/h	Control (A)	Growth rate (cell/h)	2 g (D)	Growth rate (cell/h)	4 g (E)	Growth rate (cell/h)
24	-	-	3.6±1.3	0.151	4.5±1.8	0.186
48	-	-	6.4±1.2	0.133	6.2±2.2	0.130
72	2.9±1.2	0.040	7.7±1.8	0.107	7.4±2.4	0.103
96	1.5±0.9	0.015	6.5±2.1	0.068	10.8±2.7	0.112
120	1.1±0.3	0.009	-	-	-	-
144	1.4±0.2	0.010	-	-	-	-
168	0.9±0.4	0.005	5.7±0.8	0.034	8.9±1.6	0.053
240	0.7±0.1	0.003	4.5±0.9	0.019	7.4±1.8	0.031
264	0.7±0.1	0.003	4.8±1.1	0.018	7.0±1.3	0.026

In Table 5.2, it can be observed that some data were not accessible because the time periods occurred at night and during the weekend. For instance, the cell count could not be realized at 24 h and 48 h for control, 0.175 g and 0.35 g treatments. The highest growth for the 2 g and 4 g treatments happened at 24 h, after which the algal biomass reduced in growth rate, thus, the growth rates at 120 h and 144 h were overlooked and continued with 168 h. The 4 g sample exhibited the highest growth followed by the 2 g sample, as shown in Figure 5.8. The growth resulted clearly as a function of the nutrients content for 2 g and 4 g concentrations. This implies that sufficient amount of nutrients is one of the crucial conditions required for algae growth when cultured, though it must be controlled. Also, the growth rates in the first 72 h for the 0 g, 0.175 g and 0.35 g sample treatments were as a whole higher than the rest of the durations.

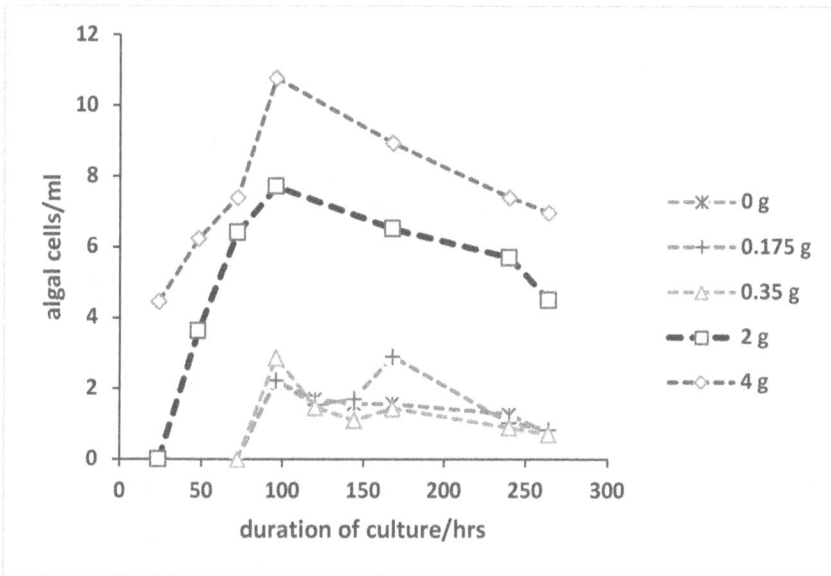

Figure 5.8 Effect of different nutrient concentrations of Yara vita fertilizer on the growth of the freshwater microalgae [39].

This may be ascribed to the fact that as the algae grew with time, the nutrients concentrations reduced, hence declining the growth rates of the algae in the successive hours. Table 5.3 illustrates the result for analyses of variance analysis (ANOVA) for various treatments. Cultures of nominally the same species often showed wide variations in cell growth rates.

Comparing the control medium with the 4 g nutrients treatment, which exhibited the best results, significant differences with respect to the cell density were observed. The P-value was less than 5% ($p < 0.05$), which is indicated by $F_{critical} < F_{observed}$ ($2.6896 < 39.4548$).

Table 5.3 ANOVA result for the data obtained during the cell count period [39]

Source of Variation	SS	df	MS	$F_{observed}$	P-value	$F_{critical}$
Between Groups	227.31	4	56.83	39.45	1.53E-11	2.69
Within Groups	43.21	30	1.44			
Total	270.52	34				

In Figure 5.9a, the growth rate of the cells increased proportionally with increased nutrients in the culture medium. The 4 g nutrient treatment exhibited significant changes when compared with the other treatments. The 2 g experimental nutrient treatment also resulted in significant differences when compared with the control, 0.35 g and 0.175 g experimental nutrient treatments. This indicated that the nutrients used in the culturing process had positive impact on the cell growth. The confidence interval for the 4 g treatment was 7.46±2.04. Figure 5.9b also shows the time taken for each of the treatments to reach their maximum growth rate. The growth rate of each of the treatments was found by dividing the mean cell count by the growth period. The maximum growth rate for each of the nutrient treatment was selected. The time taken for each of the treatment to attain their maximum growth rate was determined from the duration of the culture.

It was observed that the first three treatments reached their maximum growth after 72 h of the cultivation period, while the last two treatments reached their maximum growth within 24 h of the growth period. This depicts that the higher the nutrients concentration, the faster the growth of the freshwater green microalgae.

5.3.2 Soxhlet Extraction of Oil from Algae

15 g and 30 g dry weight of the centrifuged cultured sample and the raw sample biomass were respectively used for the extraction process. The translucent test indicated the presence of lipids in the samples suggesting that the extracted crude algal oil could subsequently be transeterified to obtain biodiesel using methanol and a catalyst. Table 5.4 shows the amount of recovered solvent and the mass of crude algal oil obtained from the green microalgae. Approximately 46.67% of hexane was recuperated from the centrifuged

sample, which was lower than the amount of hexane recuperated from the raw sample (53.33% recovery). Thus, for 150 mL of hexane

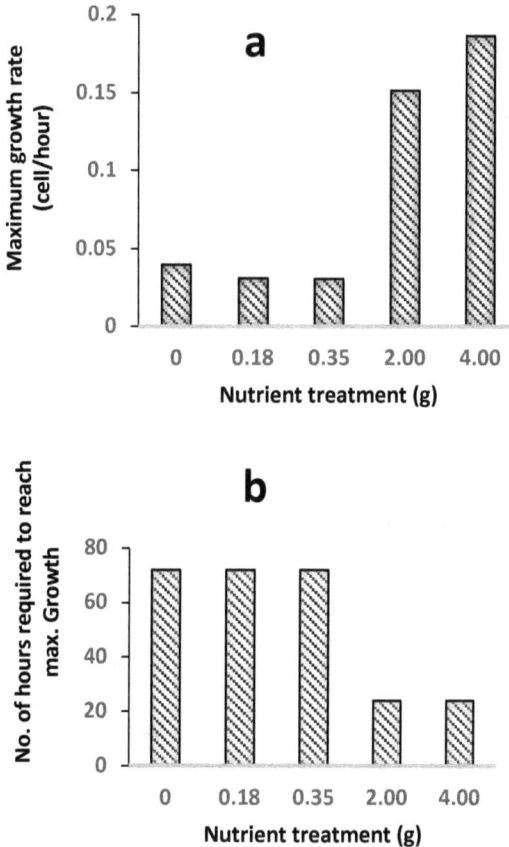

Figure 5.9 Maximum growth rate (a) and time taken (b) for nutrient treatments [39].

used for 15 g of the centrifuged sample, a recovery of 80 mL was made, whilst for 300 mL of hexane used for 30 g of the raw sample, a recovery of 140 mL was achieved. The maximum solvent recovery in the raw sample as compared with the centrifuged cultured sample may be due to the higher amount of solvent used for the raw algae, though its mass was doubled proportionally to the volume of solvent.

Figure 5.10 also shows the yield of the samples used in the extraction process. It was observed that 2.038 g and 3.826 g of oils was produced from a biomass of 15 g of the centrifuged cultured sample and 30 g of the raw sample respectively. The centrifuged culture sample had 13.6% yield, which was higher than the raw sample yield of 12.8%.

Table 5.4 Solvent recovery and the mass of obtained oil

	Centrifuged cultured sample	Raw dried sample
Initial volume of solvent	150 ml	300 ml
Solvent recovered at the end of extraction	80 ml	140 ml
Mass of algae oil obtained	2.04 g	3.83 g

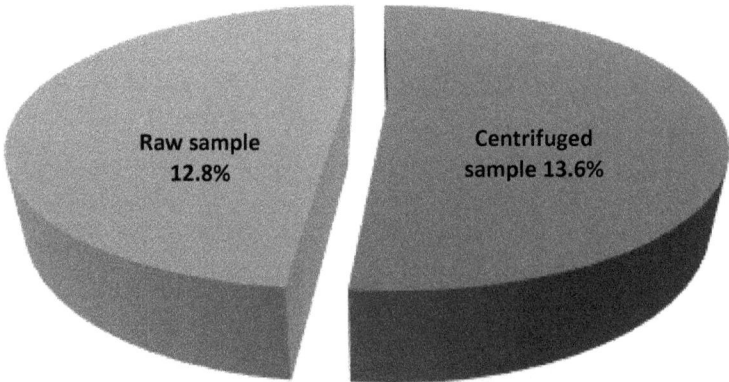

Figure 5.10 Oil yield of the centrifuged cultured and the raw sample via extractor.

This might have occurred as a result of the presence of other materials present in the raw sample, which hindered the extraction of the oil from the microalgae. However, since microalgae grow rapidly and abundantly throughout the year and the method of the oil extraction is simple, the yield can be optimized to produce higher amounts. Similar studies reported lower yields of 4.8% [30], 7.3% and 9.3% using different algae species [32].

5.3.3 Commercialization of Algal Biofuel Production

There are two main types of commercial or large-scale algal cultivation. These are open ponds and closed photobioreactor (PBR) systems. Open ponds are typically made up of open tanks and natural ponds. The green microalgae is grown in a suspension together with nutrient from fertilizer. Gas exchange is by natural contact and ambient mass transfer processes, with solar light as energy source. The maximum productivity in open pond schemes is acquired in raceway systems. Overall, there are several constraints associated with the scale-up of algal biofuel production, such as availability of carbon dioxide, supply of N and P nutrients, water constraints, energy ratio considerations, limitations of the algae biomass production technology and cost of production.

Availability of Carbon Dioxide

Carbon dioxide is important for growth of algae for biofuel production. The cultivation of each ton of algal biomass requires at least 1.83 tons of carbon dioxide [42]. Virtually all pilot plant algae cultivation depends on acquired carbon dioxide that contributes considerably (about 50%) to the cost of producing biomass. Possibly, carbon dioxide released from coal power stations can be used for the cultivation of algae, however, the amount of carbon dioxide is a major constraint for large-scale cultivation of algae.

Supply of Nitrogen and Phosphorus Nutrients

Additionally to carbon dioxide, algal growth needs nitrogen and phosphorus as key nutrients. The source of phosphorus is limited [43]. Since the sustainability of phosphorus is being challenged, there is the need to harness other ways or alternative means in which it can be derived. In the case of nitrogen, almost as much nitrogen fertilizer can be produced as needed, nevertheless, fossil energy is needed to produce the fertilizer. The amount of nitrogen and phosphorus fertilizers for agriculture is inadequate for any significant large-scale production of algal biomass cultivation for biofuel production.

Water Constraints

The supply of fresh water is insufficient to support any substantial large-scale algal biomass cultivation in any part of the world. Sources from brackish water are equally inadequate [42]. Consequently, the use of seawater and marine algae is one of the most probable options for algal cultivation for biofuel production. However, this does not eliminate completely the use of fresh water, since during the cultivation process, evaporative losses increase the salinity of the culture, thus, requiring the make-up fresh water. This option has limitations in terms of the fact that it cannot be applied to fresh water algae cultivation and would also be impossible for land lock countries and dry regions. The evaporative loss is dependent on local climatic conditions such as air temperature, wind velocity, absolute humidity and irradiance level [44].

Energy Ratio Considerations

Energy ratio is the ratio of the energy contained in the produced crude oil to the fossil energy needed to produce it. It is an important measure to determine whether the produced crude oil is meaningful or sustainable [45]. The production of biofuel from algal biomass requires contributions of energy originating from fossil fuel sources. Fossil energy is applied to pump the biomass cultivation broth and retrieve the biomass from the water by filtration and other processes as well as extract the algal crude oil from the biomass. Energy ratio of unity value means a zero net recuperation of energy from the crude oil. Preferably, energy ratio adequately above unity is highly expected, a value of at least 7 would be desired. The possible ways for achieving high-energy ratio value include the use of the biomass for biogas production, which would be subsequently used to power algal biomass cultivation. Additionally, processes such as cultivation of the biomass, recovery of the biomass from water and extraction of the crude oil should make use of the concept of process intensification, where applied fossil energy would be minimized [46]. Lui *et al.* [47] have evaluated an energy ratio of 1.4 for algal biodiesel, however, this value is very small. Thus, such energy ratio values for algal biodiesel can be greatly improved by the aforementioned measures.

Limitations of the Algae Biomass Production Technology

Though technologies for large-scale algal cultivation exist for about a decade now, the most commonly used raceway ponds have extremely low productivity as compared with what the biology of algae can provide [48]. The maximum practical productivity for algal oil of conventional raceway ponds is 37,000 L.ha^{-1}.year^{-1} [44]. Production of commercial quantities of biomass, which is economical for generating biofuels from raceway ponds is, thus, debatable [44]. On the other hand, closed cultivation systems, for example photobioreactors, are extra productive and can attain a much higher concentration of algal cells in the culture (e.g. 5 kgm^{-3}) as compared to 0.5-1.5 kgm^{-3} for raceway ponds. Nevertheless, these are expensive and need a lot of energy for operation [49].

Cost of Production

In a scenario where the resource supply (water, carbon dioxide and nutrients) is not limited, algal crude oil production would become competitive to fossil fuels. Most investors may see the initial capital involved in harnessing of crude algal oil as enormous, however, the initial capital needs in the petroleum or fossil fuels is also large and coupled with price fluctuations, which means uncertainty in investment. However, the major difficulty of producing crude algal oil is the large quantity of biomass for sustainable and low cost production. Chisti [44] has estimated that the algal biomass with oil content 40% weight will have to be produced at a cost of not more than $0.25 per kg for it to compete with fossil fuels at around the price of $629 per cubic meter ($100 per barrel). The actual cost of cultivating the biomass currently appears to be 10 fold greater [50]. Process intensification of the production technology and operations at a more economic scale of 200 tons per year are estimated to reduce the cost of producing the dry biomass to €12.6 per kg [51]. In a scenario where lower cost raceway pond based production facility is used, it can produce the biomass at 10% of this price. However, the biomass will still be nearly 5-fold more expensive compared to the maximum acceptable production price of $0.25 per kg [44] required to attain the cost effectiveness with fossil fuel derived fuels.

Carbon Footprint

Carbon footprint is a measure of the amount of carbon liberated in production and use of a given quantity of material [52]. The carbon footprint of crude algal oil relative to the carbon footprint of the petroleum fuel it displaces is a key factor in establishing whether its production is meaningful. The carbon footprint of an acceptable biofuel must be smaller than the carbon footprint of petroleum on an equal energy balance. Shirvani *et al.* [53] reported life cycle energy analysis for greenhouse gas emissions for algal biodiesel which ranged from 78 to 351 g.MJ^{-1}, depending on how the electricity is applied in the production of algal fuel [53]. For comparison, the greenhouse gas emission for petroleum diesel of 82.3 g.MJ^{-1} of energy content has been reported [54].

General Solution to the Impediments on Commercialization of Algal Fuels

Availability of Carbon Dioxide
For future prospects, there should be the possibility of considering the usage of carbon dioxide capture as the source for algae culturing. Unfortunately, the exhaust fumes from the automobiles cannot be directly used for the growth of algae, since the exhausts contain other gaseous components apart from carbon dioxide. Also, concentrated sources of carbon dioxide are largely the flue gases formed during power generation from combustion of coal. Furthermore, the cement industry generates concentrated source of CO_2. These concentrated sources can be exploited for the low cost algal cultivation.

Supply of Nitrogen and Phosphorus Nutrients
Herein, there is the need to find alternative ways to provide the necessary nutrients for the algal biomass cultivation. The most appropriate alternative is the application of sewage from densely populated cities and metropolis for the algae cultivation. This is recommended as it comprises of both nitrogen and phosphorus and, hence, mimics artificial fertilizer, thus, making it suitable for algae culture. There exists the possibility of recycling human urine, manure from animals and bones of animals into phosphorus.

Water Constraints
Wastewater can be employed, as large volumes are generated from densely populated communities with nutrients needed for algal cultivation. There is a need to introduce the concept of process intensification where it is necessary to reuse/recycle and reduce the amount of water needed for the process. Moreover, the usage of smaller bioreactors is encouraged for the process of algal cultivation to reduce the amount of water used.

Energy Ratio Considerations
A promising way for achieving high energy ratio is to transform the

biomass into biogas production, which would be subsequently used to power algal biomass cultivation. Furthermore, processes such as cultivation of algal biomass, recovery of biomass from water and extraction of crude oil should make use of the concept of process intensification, where applied fossil energy would be minimized as well as integrated [46]. The energy efficiency, in addition to biomass productivity, of raceway ponds can be enhanced to some extent by changes made to the design [46]. Moreover, analogous improvement in energy efficiency has also been envisioned in photobioreactors [55].

Limitations of the Algae Biomass Production Technology
Innovative biomass production methods, which are economical and energy efficient and depend on sunlight, are required to achieve high productivity and biomass concentration in the culture. Hence, the introduction of shallow raceway ponds or other low cost systems such as continuous stirred tank bioreactors and photobioreactors with high surface to volume ratio to maximize sunlight capture and transmission into the algal culture may be considered to overcome the limitations [50].

Cost of Production
Currently, most commercial algal biofuel production systems are expensive. To make the current processes more economical, there is a need to introduce inexpensive and low energy processes for algal biomass cultivation, biomass recovery and oil extraction. Furthermore, the introduction of the concept of process intensification of the production technology and operation at a more economic scale of 200 tons per year are estimated to reduce the cost of producing the dry biomass to € 12.6 per kg [51].

Carbon Footprint
Nitrous oxide, a greenhouse gas which is much more potent than carbon dioxide, has been occasionally detected in microalgae cultivation systems. Apparently, nitrates produce nitrous oxide by denitrifying bacteria, which contaminates the culture [56]. Thus, in a continuously mixed open ponds and aerated photobioreactors, sufficient oxygen levels even during the dark are needed to prevent significant emissions of nitrous oxide. A nitrous oxide emission level of 0.002% of the nitrogen fertilizer used has been estimated for toxic algal cultures [56].

5.3.4 Technology Transfer of the Batch Reactor Case Study into CSTR System

For the future prospects of the batch reactor case study to pave the way for CSTR system, there have been comprehensive assessments of different large-scale systems for algal cultivation of biomass for biofuel production. A detailed summary of the large-scale algal biomass culture systems with various considered parameters is outlined in Table 5.1 [38]. It has been observed that tubular bioreactors and CSTR systems are good candidates for large-scale algal biomass cultivation. Moreover, Borowitzka [38] reported that CSTR scaleup is difficult, however, it does not imply that it cannot be realized. Hence, to realize a successful implementation of scaleup with CSTR, there is a need to modify the reactor regime to take care of the challenges. The CSTR that would be employed must mimic the modified batch bioreactor, which is depicted in Figure 5.3. The introduction of the modified CSTR would make the scaleup process feasible, as hydrodynamic issues and viscous flow problems would be solved.

Furthermore, there is a need to implement the solution proposed in this work to the impediments mentioned in earlier sections. As per the suggested solution, photobioreactors were recommended as the appropriate system to achieve commercialization. For instance, in a continuously mixed open ponds and aerated photobioreactors, there is usually sufficient oxygen even during the dark to prevent significant amount of nitrous oxide emissions. Thus, the introduction of shallow raceway ponds or other low cost systems such as continuous stirred tank bioreactors and photobioreactors with high surface to volume ratio to maximize sunlight capture and transmission into the algal culture should be considered [50]. This is introduced to improve sunlight penetration into the algal biomass cultivation systems to enhance growth.

5.4 Conclusions

The first part of this work has studied the production of biodiesel from green microalgae as a batch case study. The second part of the study has outlined the challenges and various solutions associated with the commercialization of algal biofuel.

The obtained results revealed that the cultured algae treated with 4 g Yara vita nutrient had the maximum growth of 1.186 cell/h, whereas the 2 g treatment exhibited the second highest growth rate, both of which attaining growth in the first 24 h. However, the lesser extent of treatment, such as 0 g, 0.175 g and 0.35 g, attained their highest growth of 0.04 cell/h, 0.031 cell/h and 0.031 cell/h respectively in a period of 72 h. This suggested that the greater the nutrients concentration, the faster was the growth of the green microalgae. Additionally, 0.038 g of crude oil was obtained from 15 g of the centrifuged cultured green microalgae with a yield of 13.6%. However, 3.83 g of the oil was acquired from 30 g of the raw algae sample and with a yield of 12.8%. The result confirmed that for current processes to be more economical, there is a need to introduce low cost and low energy processes for algal biomass cultivation, biomass recovery and oil extraction. The lower yield in the raw sample compared to the centrifuged sample could be attributed to the presence of impurities, which might have hindered the oil extraction.

Although commercialization of algal biomass for fuel has appreciable number of impediments, it is still possible to achieve success by implementing alternative solutions. The first step for selecting a large-scale algal cultivation method relies on the appropriate reactor regime or type. From Table 5.1, it can be clearly observed that tubular bioreactors and CSTR systems are the most feasible systems for large-scale algal biomass cultivation. Sterilization process should be enhanced in large-scale algal biomass cultivation for biofuel generation, as evidently nitrates produce nitrous oxide by denitrifying bacteria, which contaminates the culture [56].

The results of the study demonstrate that biodiesel, which is more environmentally friendly than fossil diesels, can be produced from the plentiful and sustainable green microalgae using hexane with overall lower extent of carbon dioxide emission. Moreover, since algae grow easily and copiously throughout the year, acquiring a yield of 12.8% or 13.6% is quite substantial.

Acknowledgement

The author is grateful to the University of Cape Coast, Ghana for financial support. The author thanks Mr. Siisu Salifu and Mr. Francis Kotoka of the University of Cape Coast for experimental assistance.

References

1. Shay, E. G. (1993) Diesel fuel from vegetable oils: status and oppor-

tunities. *Biomass and Bioenergy*, **4**(4), 227-242.

2. Rittmann, B. E. (2008) Opportunities for renewable bioenergy using microorganisms. *Biotechnology and Bioengineering*, **100**, 203-212.

3. Ryckebosch, E., Muylaert, K., and Foubert, I. (2012) Optimization of an analytical procedure for extraction of lipids from microalgae. *Journal of the American Oil Chemists' Society*, **89**(2), 189-198.

4. Enweremadu, C. C., and Alamu, O. J. (2010) Development and characterization of biodiesel from Nigerian shea nut butter (*Vitellaria paradoxa*). *International Agrophysics*, **24**(1), 29-34.

5. Schenk, P. M., Thomas-Hall, S. R., Stephens, E., Marx, U. C., Mussgnug, J. H., Posten, C., Kruse, O., and Hankamer, B. (2008) Second generation biofuels: high-efficiency microalgae for biodiesel production. *BioEnergy Research*, **1**, 20-43.

6. Lee, J. Y., Jun, S. Y., Ahn, C. Y., and Oh, H. M. (2009) Comparison of several methods for effective lipid extraction from microalgae. *Bioresource Technology*, **101**(1), S75-S77.

7. Wang, L., Li, Y., Chen, P., Min, M., Chen, Y., Zhu, J., and Ruan, R. R. (2010) Anaerobic digested dairy manure as a nutrient supplement for cultivation of oil-rich green microalgae Chlorella sp. *Bioresource Technology*, **101**(8), 2623-2628.

8. Xue, F., Zhang, X., Luo, H., and Tan, T. (2006) A new method for preparing raw material for biodiesel production. *Process Biochemistry*, **41**(7), 1699-1702.

9. Tran, H. L., Hong, S. J., and Lee, C. G. (2009) Evaluation of extraction methods for recovery of fatty acids from Botryococcus braunii LB 572 and Synechocystis sp. PCC 6803. *Biotechnology and Bioprocess Engineering*, **14**(2), 187-192.

10. McNichol, J., MacDougall, K. M., Melanson, J. E., and McGinn, P. J. (2012). Suitability of soxhlet extraction to quantify microalgal fatty acids as determined by comparison with in situ transesterification. *Lipids*, **47**(2), 195-207.

11. Borowitzka, M. A. (1988) Fats, oils and hydrocarbons. In: Micro-Algal Biotechnology, Borowitzka, M. A., and Borowitzka, L. J. (eds.), Cambridge University Press, UK, pp. 257-287.

12. Chisti, Y. (2007) Biodiesel from microalgae. *Biotechnology Advances*, **25**, 294-306.

13. Spoehr, H. A., and Milner, H. W. (1949) The chemical composition of Chlorella, effect of environmental conditions. *Plant Physiology*, **24**, 120-149.

14. Meher, L. C., Vidya Sagar, D., and Naik, S. N. (2006). Technical aspects of biodiesel production by transesterification-A review. *Renewable and Sustainable Energy Reviews*, **10**, 248-268.

15. Field, C., Behrenfeld, M., Randerson, J., Falkowski, P. (1998) Primary production of the biosphere: Integrating terrestrial and oceanic

components. *Science,* **281**, 237-240.

16. Milano, J., Org, H. C., Masjuki, H. H., Chong, W. T., Lam, M. K., Loh, P. K., and Vellayan, V. (2016) Microalgae biofuels as an alternative to fossil fuel for power generation. *Renewable and Sustainable Energy Reviews*, **58**, 180-197.

17. Sayre, R. (2010) Microalgae: the potential for carbon capture. *BioScience*, **60**, 722-727.

18. Heller, W. P., Kissinger, K. R., Matsumoto, T. K., and Keith, L. M. (2015) Utilization of papaya waste and oil production by Chlorella protothecoides. *Algal Research*, **12**, 156-160.

19. Liang, Y., Osada, K., Sunaga, Y., Yoshino, T., Bowler, C., and Tanaka, T. (2015) Dynamic oil body generation in the marine oleaginous Fistulifera solaris in response to nutrient limitation as revealed by morphological and lipidomic analysis. *Algal Research*, **12**, 359-367.

20. Kamalanathan, M., Gleadow, R., and Beardall, J. (2015) Impact of phosphorus availability on lipid production by Chlamydomonas reinhardtii. *Algal Research*, **12**, 191-196.

21. Valdez-Ojeda, R., Gonzalez-Munoz, M., Us-Vazquez, R., Narvaez-Zapata, J., Chavarria-Hernandez, J. C., Lopez-Adrian, S., Barahona-Perez, F., Toledano-Thompson, T., Gardduno-Solorzano, G., and Medrano Escobedo-Gracia, R. M. (2015) Characterization of five fresh water microalgae with potential for biodiesel production. *Algal Research*, **7**, 33-44.

22. Mubarak, M., Shaija, A., and Sunchithra, T. V. (2015) A review of the extraction of lipid from microalgae for biodiesel production. *Algal Research*, **7**, 117-123.

23. Halim, R., Harun, R., Danquah, M. K., and Webley, P. A. (2012) Microalgal cell disruption for biofuel development. *Applied Energy*, **91**, 116-121.

24. Roger, J. N., Rosenberg, J. N., Guzman, B. J., Oh, V. H., Mimbela, L. E., Ghassemi, A., Betenbaugh, M. J., Oyler, G. A., and Donohue, M. D. (2014) A critical analysis of puddlewheel driven raceway ponds for algal biofuel production at commercial scales. *Algal Research*, **4**, 76-88.

25. Bigelow, N., Barker, J., Ryken, S., Patterson, J., Hardin, W., Barlow, S., Deodato, C., and Cattolico, R. A. (2013) Chrysochromulina sp.: A proposed lipid standard for the algal biofuel industry and its application to diverse taxa for screening lipid content. *Algal Research*, **2**, 385-393.

26. Sharma, T., Gour, R. S., Kant, A., and Chauhan, R. S., (2015) Lipid content in Scenedesmus species correlates with multiple genes of fatty acid and triacylglycerol biosynthetic pathways. *Algal Research*, **12**, 341-349.

27. Chisti, Y. (2008) Biodiesel from microalgae beats bioethanol. *Trends in Biotechnology*, **26**, 126-131.

28. Folch, J., Lees, M., and Sloane, S. G. H. (1957) A simple method for the isolation and purification of total lipids from animal tissues. *The Journal of Biological Chemistry,* **226**, 497-509.

29. Bligh, E. G., and Dyer, W. J. (1959) A rapid method of total lipid extraction and purification. *Canadian Journal of Biochemistry and Physiology,* **37**, 911-917.

30. Ramanathan, R. K., Polur, H. R., and Muthu, A. (2015). Lipid extraction methods from microalgae: a comprehensive review. *Frontiers in Energy Research,* doi: 10.3389/fenrg.2014.00061.

31. Sheng, J., Vannela, R., and Rittmann, B. E. (2011) Evaluation of methods to extract and quantify lipids from *Synechocystis* PCC 6803. *Bioresource Technology,* **102**, 1697-1703.

32. Hossain, A. S., Salleh, A., Boyce, A. N., Chowdhury, P., and Naqiuddin, M. (2008) Biodiesel fuel production from algae as renewable energy. *American Journal of Biochemistry and Biotechnology,* **4**(3), 250-254.

33. Balasubramanian, S., Allen, J. D., Kanitkar, A., and Boldor, D. (2010) Oil extraction from Scenedesmus obliquus using a continuous microwave system – design, optimization, and quality characterization. *Bioresource Technology,* **102**, 3396-3403.

34. Miron, A. S., Garcia, M. C. C., Gomez, A. C., Camacho, F. G., Grima, E. M., and Chisti, Y. (2003) Shear stress tolerance and biochemical characterization of Phaeodactylum tricornutum in quasi steady-state continuous culture in outdoor photobioreactors. *Biochemical Engineering Journal,* **16**, 287-297.

35. Belarbi, E. H., Milona, E., and Chisti, Y. (2000) A process for high yield and scaleable recovery of high purity eicosapentaenoic acid esters from microalgae and fish oil. *Enzyme and Microbial Technology,* **26**, 516-529.

36. Stein, J. R. (1973) *Handbook for Phycological Methods: Culture Methods and Growth Measurements,* Cambridge University Press, UK.

37. Green, D. W., and Perry, R. H. (1999) *Perry's Chemical Engineers' Handbook,* 8th edition, McGraw-Hill, USA.

38. Borowitzka, M. A. (1999) Commercial production of microalgae: ponds, tanks, tubes and fermenters. *Journal of Biotechnology,* **70**, 313-321.

39. Tulashie, S. K., and Salifu, S. (2017) Potential production of biodiesel from green microalgae. *Biofuels,* DOI: 10.1080/17597269. 2017.1348188.

40. Fritch, F. E. (1945) *The Structure and Reproduction of the Algae. I and II,* Cambridge University Press, UK.

41. Browne, B., Gibbs, R., McLeod, J., Parker, M., Schwanda, W., and Warren, K. (2009) Oil Extraction from Microalgae. *Algae Oil Extraction Capstone.* Online: http://mickpeterson.org/Classes/Design/

2009_10/Projects/Algae/AlgaeFinalReport.pdf (assessed 19th June 2018).

42. Chisti, Y. (2007) Constraints to commercialization of algal fuels. *Journal of Biotechnology*, **167**, 201-214.

43. Cordell, D., Drangert, J.-O., and White, S. (2009) The story of phosphorus: global food security and food for thought. *Global Environmental Change*, **19**, 292-305.

44. Chisti, Y. (2012) Raceways based production of algal crude oil. In: *Microalgal Biotechnology: Potential and Production*, Posten, C., and Walter, C. (eds.), de Gruyter, Germany, pp. 113-146.

45. Chisti, Y. (2008b) Response to rejoinders: do biofuel from microalgae beat biofuels from terrestrial plants? *Trends in Biotechnology*, **26**, 351-352.

46. Sompech, K., Chisti, Y., and Srinophakun, T. (2012) Design of raceway pond for the production of microalgae. *Biofuels*, **3**, 387-397.

47. Lui, X., Clarens, A. F., and Colosi, L. M. (2012a) Algae biodiesel has potential despite inconclusive results to date. *Bioresource Technology*, **104**, 803-806.

48. Terry, K. L., and Raymond, L. P. (1985) System design for the autotrophic production of microalgae. *Enzyme and Microbial Technology*, **7**, 474-487.

49. Wang, B., Lan, C. Q., and Horsman, M. (2012) Closed photobioreactors for production of microalgal biomasses. *Biotechnology Advances*, **30**, 904-912.

50. Chisti, Y. (2013) Constraints to commercialization of algal fuels. *Journal of Biotechnology*, **167**, 201-214.

51. Acien, F. G., Fernandez, J. M., Magan, J. J., and Molina, E. (2012) Production cost of a real microalgae production plant and strategies to reduce it. *Biotechnology Advances*, **30**, 1344-1353.

52. Zamboni, A., Murphy, R. J., Woods, J., Bezzo, F., and Shah, N. (2011) Biofuels carbon footprint: whole-systems optimization for GHG emissions reduction. *Bioresource Technology*, **102**, 7457-7465.

53. Shirvani, T., Yan, X.-Y., Inderwildi, O. R., Edwards, P. P., and King, D. A. (2011) Life cycle energy for greenhouse gas analysis for algal-derived biodiesel. *Energy and Environmental Science*, **4**, 3773-3778.

54. Hill, J., Nelson, E., Tilman, D., Polasky, S., and Tiffany, D. (2006) Environmental, economic and energetic costs and benefits of biodiesel and ethanol biofuels. *Proceedings of the National Academy of Sciences of United State of America*, **103**, 11206-11210.

55. Wongluang, P., Chisti, Y., and Srinophakun, T. (2013) Optimal hydrodynamic design of tubular photobioreactors. *Journal of Chemical Technology and Biotechnology*, **88**, 55-61.

56. Fagerstone, K. D., Quinn, J. C., Bradley, T. H., De Long, S. K., Marchese, A. J. (2011) Quantitative measurement of direct nitrous oxide

emissions from microalgae cultivation. *Environmental Science and Technology*, **45**, 9449-9456.

Chapter 6

Advancements in the Commercialization of Lignocellulosic Bioethanol

Saumita Banerjee,[1] Sandeep Mudliar[2] and Ramkrishna Sen[1,*]
[1]Indian Institute of Technology Kharagpur, Kharagpur, West Bengal, India
[2]Central Food Technological Research Institute, Mysore, Karnataka, India
*Corresponding author: rksen@yahoo.com

6.1 Introduction

Bioethanol is a well-proven, viable alternative to petroleum based fuels, especially for the transportation sector. The multiple advantages of supplementing/replacing gasoline with bioethanol, viz. greenhouse gas reduction, energy independence, resulting foreign exchange savings, employment opportunities, etc., are mitigated mainly due to the limited availability of conventional starch and sugar sources and the very real 'food-vs-fuel' dilemma [1]. A way out of this problem and a win-win for all stakeholders is to develop the technology to convert lignocellulosic biomass (agricultural and forest residues, industrial and municipal waste, energy crops, etc.) into their component sugars and fermenting these into bioethanol (and many other bio-products). In addition to the aforementioned advantages of bioethanol as fuel-substitute, lignocellulosic bioethanol offers some additional benefits: efficient and sustainable use of biomass otherwise getting wasted (or worse, creating significant air pollution due to incorrect disposal methods) and a seamless bridging of the gap between agronomy and industries. Many important industrial chemicals (e.g. acetic acid, citric acid, lactic acid, xylitol, sorbitol, isopropanol, oxalic acid and many others) and end products (e.g. adhesives, cosmetics, food additives, pharmaceuticals, solvents, surfactants, etc.) can also be bio-produced by the action of microbes on key sugars such as glucose and xylose - the main component sugars in lignocellulose. One may, thus, witness the emergence of a true biorefinery [2]. The last decade has seen tremendous progress in research efforts to usher in new technologies for the conversion of lignocellulose into component sugars and ultimately into fermentation products such as

Biofuels, edited by Vikas Mittal

bioethanol. The research efforts are now beginning to bear fruit, as we are witnessing a steady growth in the number and scale of biorefineries capable of economically and sustainably converting lignocellulosic biomass into bioethanol (and other bioproducts). This chapter takes a look at some of these technologies that are at the forefront of what might be called an "agro-industrial revolution" in its true sense.

6.2 Notable Large-scale or Near-commercial Technologies for Lignocellulosic Bioethanol Production

6.2.1 Mineral Acid Based Biomass Hydrolysis and Microbial Fermentation

Arkenol® Process

Arkenol constructs and operates biorefineries globally to produce a variety of bio-based chemicals and transportation fuels [3]. The Arkenol biorefineries utilize a proprietary technology, called "concentrated acid hydrolysis", to convert biomass in various forms into simple sugars. The sugars are then fermented/chemically converted into a variety of fuels and chemicals, ranging from beverage alcohol and fuel-ethanol to citric acid and xantham gum for food uses. Arkenol has developed significant proprietary improvements to the well-known concentrated acid hydrolysis such that the process of biomass conversion into sugars is economically viable and is ready for commercial implementation.

To demonstrate the efficacy of the technology, Arkenol has constructed a pilot plant in Orange, California. An integrated, full-scale commercial process plant consists of five basic unit operations:
- Feedstock preparation
- Decrystallization/hydrolysis reaction vessel
- Solids/liquid filtration
- Separation of the acid and sugars
- Fermentation of the sugars and
- Product purification

BlueFire Renewables [4] is the exclusive North American licensee of the Arkenol technology. BlueFire mentions to be the only viable cellulose conversion company with demonstrated production experience from wood wastes, urban trash (post-sorted MSW), rice and wheat straws and other agricultural residues. The company mentions to have full-scale bioethanol projects currently in various stages of development.

The Arkenol technology is licensed to GS Caltex in Korea and JGC Corporation in Japan and SE Asia. Since 2003, the technology has been successfully used in the IZUMI pilot plant operated by JGC, to produce ethanol for the Japanese transportation fuel market.

6.2.2 Thermo-chemical Pre-treatment of Biomass, Enzyme Hydrolysis and Microbial Fermentation

Clariant Sunliquid® Process

Clariant, a leading specialty chemicals company, has developed the sunliquid® technology, a technically and economically efficient process for converting agricultural residues into second generation biofuels [5]. The technology is built on established process technology, with a completely integrated process design. It features innovative concepts such as the integrated production of feedstock and process specific enzymes, simultaneous C5 and C6 fermentation and an energy saving method for product purification.

The sunliquid® process has been running successfully in a pilot plant at Clariant's research center in Munich since early 2009. A demonstration plant in Straubing near Munich produces up to 1,000 tons of bioethanol per year [6]. Clariant is now setting up a first-of-its kind large-scale sunliquid® production plant, with funding from the European Union's Seventh Framework Programme (FP7).

Clariant has entered into a license agreement with Enviral, the largest producer of bioethanol in Slovakia, to design, build and operate a new full scale commercial cellulosic ethanol plant at its Leopoldov site with Clariant's sunliquid® technology [7]. This plant is planned to be integrated into the existing facilities at Enviral's Leopoldov site in Slovakia with an annual production of 50,000 tons.

This technology is being successfully used in the production of isobutene from wheat straw hydrolysate, produced at Clariant's Straubing facility [8]. The project is a collaboration with Global Bioenergies, a European company developing a process to convert renewable resources into hydrocarbons through fermentation.

Beta Renewables' PROESA® Process

Beta Renewables' PROESA™ process is another promising technology that combines thermo-chemical biomass pre-treatment and hybrid enzymatic hydrolysis and fermentation by the SSCF process (simultaneous saccharification and co-fermentation) [9]. Beta Renewables has the credit of starting up the first plant in the world for the industrial production of second-generation bioethanol [10]. At full capacity, the plant produces 40,000 tons of bioethanol per year. The plant uses 270,000 t/y (at maximum potential) of biomass (rice straw, wheat straw and Arundo Donax, the common giant reed) and produces 13 MW of electricity, entirely produced using lignin, making the plant totally self-sufficient with regards to its energy consumption. The plant is in successful operation since 2013.

GranBio's Bioflex 1 industrial unit is a commercial-scale cellulosic ethanol factory in Brazil, operational since September 2014 [11]. It has capacity to produce 82 million liters of biofuel per year. The plant is based on the PROSEA technology. Beta Renewables has many other commercial scale projects under construction.

Inbicon Process

A second-generation ethanol demonstration plant was opened in Kalundborg, Denmark in 2009 by Inbicon [12]. The Kalundborg biorefinery plant makes use of the heat energy (steam) generated by a nearby coal-fired power plant for ethanol production [13]. The plant annually produces 1.5 million gallons (5.4 million litres) of ethanol, 13,000 t of powdered lignin pellets (used as a renewable fuel), natural bacteria inhibitor and 11,100 t of C5 molasses, which can be further used for bioethanol or biogas production. It converts wheat straw into ethanol.

The core technology involves steam-based pre-treatment of biomass, followed by simultaneous enzymatic hydrolysis and microbial fermentation. The uniqueness of the process is that it operates at very high dry matter concentrations (25-40% DM), yielding a high ethanol concentration of approx. 10% (vol/vol) in the fermentation liquid. Fermentation is carried out in two parts: first in a horizontal reactor to handle the highly viscous fibers and then pumped to standard vertical fermenters. From the biomass preparation step to the first fermentation step (in horizontal reactors), the Inbicon process is in continuous operation, with a residence time of 30-40 hours. The vertical fermenters are run in batch mode with a process time of 100 hours.

The Inbicon technology is ready for full commercial scale deployment. Maabjerg Energy Center in Denmark, a project incorporating a full scale cellulosic ethanol plant with a capacity of 80 million litres per year is based on Inbicon's technology and is expected to begin production from 2018 [14].

6.2.3 Biomass Fractionation, Enzyme Hydrolysis and Microbial Fermentation

DBT-ICT Process

Institute of Chemical Technology (ICT) in India has set up a cellulosic ethanol demonstration plant based on in-house developed technology [15]. The plant is in operation for the last two years and is capable of producing 10 tons/day. The technology can process both woody and non-woody type biomass and can convert biomass into ethanol in less than 24 hours. The process involves a two-step fractionation process to obtain a cellulose slurry, which is then hydrolyzed and fermented into ethanol by a fed-batch process. The enzymes for hydrolysis are recycled by membrane filtration. Xylose and lignin are recovered for processing and further uses. The process is designed as a continuous system.

ICT is collaborating with L&T Hydrocarbon Engineering, a leading provider of design-to-build engineering and construction solutions, to offer complete solutions in setting up 2G ethanol plants in terms of process license, technology knowhow, basic engineering, EPC and EPCM services [16]. The technology is mentioned to provide one of the lowest project life costs.

ICT is providing its technology know-how to a biorefinery featuring a commercial scale ethanol plant being developed by Hindustan Petroleum Development Corporation. The ethanol plant will be capable of producing 100 kilo liters of ethanol per day [17].

Arbiom® Process

Arbiom Inc., a company located in the Research Triangle Park, North Carolina, works on a different business strategy [18]. It has the technology to convert biomass into its component sugars and lignin. It then works with its partners to design the process to convert biomass into sugars and then into partner products.

Arbiom's process uses a low-temperature "biocompatible" process to unlock the biomass structure. The process involves dissolving the cellulose in biomass, using an appropriate cellulose solvent, such as phosphoric acid or an ionic liquid, followed by multi-stage hydrolysis of the dissolved cellulose using enzymes. The solvents are recycled and reused.

Arbiom's process operations center in Norton, Virginia has a nameplate capacity of 1 ton/day of biomass [19]. The pilot facility uses a wide range of feedstocks, from hardwoods to agricultural wastes, for production of C5 and C6 sugars and lignin.

Arbiom is partnering with many companies to scale up the process to commercial levels. A notable partnership is with Norske Skog Golbey, a pulp and paper company, on a biorefinery project named "Bioskog" [20]. The commercial scale biorefinery will have a planned processing capacity of 40,000 tons of biomass per year and will be operational by 2018.

6.3 Conclusions/Outlook

Second generation bioethanol, produced from waste biomass, is racing toward commercialization and it may not be long before biomass will not be "waste". Since biomass availability is fairly consistent across the world, the phenomenon will spread on a global scale and the energy monopoly enjoyed by oil-producing nations will see a decline. The process of lignocellulose to ethanol bioconversion has been proven to be effective in reducing green-house gas emissions. This will be an important step toward climate change mitigation.

The rise of the technology from research-to-commercial scale has not been without problems though. There have been several cases of failure even at commercial scale implementations. DuPont celebrated opening of its cellulosic biofuel facility in Nevada, Iowa, in 2015 [21], however, two years later decided to sell the company's cellulosic ethanol facility [22]. Abengoa Bioenergy Biomass of Kansas sold its Hugoton cellulosic ethanol plant having a 25 million gallon per year nameplate capacity to Synata Bio [23]. The construction of BlueFire's 19 million gallons of ethanol per year Fulton plant did not complete [24]. This being said, the success achieved so far is also commendable. The list of commercial/near-commercial cellulosic ethanol plants listed here is not exhaustive and there are many more plants that are successfully operating today, with elaborate plans of further scale-up in terms of both capacity and number. Technologies by POET-DSM [25], Enerkem [26], ST1 Biofuels [27] and many others are being implemented at demonstration- and commercial-scales. Furthermore, new technologies are also being evaluated right from the lab scale and upward. Solid acid catalysts capable of hydrolyzing the biomass without the use of enzymes is one such example [28]. The point here is to unlock the tightly bound structure of lignocellulose in as simple manner as possible. Any means to accomplish this will drive the growth in future.

The general consensus seems to be to exercise extreme caution with respect to the actual implementation of such commercial scale projects. Successful pilot-scale or demonstration-scale performance may not be a guarantee for success at a commercial scale, simply because the actual process is much more than simple economics. Biomass availability, its price fluctuations, costs incurred in enzyme procurement and the price fluctuations of crude oil can contribute to the uncertainty associated with the success of cellulosic ethanol. The smart way out is to branch out these factors as much as possible - using a technology that can handle all kinds of biomass (to reduce biomass dependence), produce enzymes on-site (or perhaps use a technology that does not use enzymes at all - like solid acid catalysts) and if possible, co-locate the production of a high-value bio-product with the ethanol plant (to fight a crude-oil price challenge). Robust technology, combined with smart implementation, comprises the "secret sauce".

References

1. Ibeto, C. N., Ofoefule, A. U., and Agbo, K. E. (2011) A global overview of biomass potentials for bioethanol production: A renewable alternative fuel. *Trends in Applied Sciences Research*, **6**(5), 410-425.
2. Banerjee, S., Mudliar, S., Sen, R., Giri, B., Satpute, D., Chakrabarti, T., and Pandey, R. A. (2010) Commercializing lignocellulosic bioethanol: technology bottlenecks and possible remedies. *Biofuels, Bioproducts & Biorefining*, **4**, 77-93.
3. Our Technology (2014) *Arkenol*, USA. Online: http://arkenol.com/technology/ (assessed 16th May 2018).
4. BlueFire Approach (2011) *BlueFire Renewables*, USA. Online: www.bfreinc.com (assessed 16th May 2018).
5. Sunliquid® FP7 (2018) *Clariant Produkte (Deutschland) GmbH*, Germany. Online: https://www.sunliquid-project-fp7.eu (assessed 16th May 2018).
6. Cellulosic Ethanol from Agricultural Residues - Think Ahead, Think Sunliquid (2018) *Clariant*, Germany. Online: https://www.clariant.com/sunliquid (assessed 16th May 2018).
7. News: Clariant and Enviral Announce First License Agreement on

Sunliquid® Cellulosic Ethanol Technology (September 2017) *Clariant*, Germany. Online: https://www.clariant.com/en/Corporate/News/2017/09/Clariant-and-Enviral-announce-first-license-agreement-on-sunliquid-cellulosic-ethanol-technology (assessed 16th May 2018).

8. News: Global Bioenergies and Clariant Announce the First Isobutene Production from a Wheat Straw Hydrolysate (October 2016) *SynBioBeta*, USA. Online: https://synbiobeta.com/global-bioenergies-clariant-announce-first-isobutene-production-wheat-straw-hydrolysate/ (assessed 16th May 2018).

9. PROESA (2018) *Beta Renewables*, Italy. Online: http://www.beta-renewables.com/en/proesa/what-is-it (assessed 16th May 2018).

10. News: World's First Advanced Biofuels Facility Opens (October 2013) *Novozymes*, Denmark. Online: https://www.novozymes.com/en/news/news-archive/2013/10/worlds-first-advanced-biofuels-facility-opens (assessed 16th May 2018).

11. Bioflex I (2018) *GranBio*, Brazil. Online: http://www.granbio.com.br/en/conteudos/biofuels/ (assessed 16th May 2018).

12. Kalundborg Bioethanol Demonstration Plant (2018) *Chemicals Technology*. Online: https://www.chemicals-technology.com/projects/kalundborg_bioethano/ (assessed 16th May 2018).

13. KACELLE: Bringing Cellulosic Ethanol to Industrial Production at Kalundborg, Denmark (April 2014). Online: https://biorefiningalliance.com/wp-content/uploads/2014/12/KACELLE-PUBLISHABLE-SUMMARY-web.pdf (assessed 16th May 2018).

14. Maabjerg Energy Center is a Bio-energy Plant of Ground-breaking Dimensions (2018) Maabjerg Energy Center, Denmark. Online: https://www.maabjergenergycenter.com/ (assessed 16th May 2018).

15. Lali, A. (2018) DBT-ICT Technology Platforms For Advanced Biofuels. *EU-India Conference on Advanced Biofuels*, India. Online: https://ec.europa.eu/energy/sites/ener/files/documents/13_arvind_lali-dbt-ict.pdf (accessed 16th May, 2018).

16. News: Larsen Arm Ties Up with ICT to Build Ethanol Plants (May 2017) *The Hindu, Business Line*, India. Online: https://www.thehindubusinessline.com/companies/larsen-arm-ties-up-with-ict-to-build-ethanol-plants/article9713218.ece (assessed 16th May 2018).

17. News: HPCL to Set Up Rs 600-Crore Bio-ethanol Unit in Bathinda (May 2017) *The Tribune*, India. Online: http://www.tribuneindia.com/news/business/hpcl-to-set-up-rs-600-crore-bio-ethanol-unit-in-bathinda/408603.html (assessed 16th May 2018).

18. Wood to Food (2018) *Arbiom*, USA. Online: http://www.arbiom.com/ (assessed 16th May 2018).

19. Arbiom Technology (2018) *Arbiom*, USA. Online: http://www.ar-biom.com/commercialization/process-operations/ (assessed 16th May 2018).
20. News: Arbiom Announces Bioskog, A Biorefinery Project in Golbey, France in Partnership with Norske Skog Golbey (November 2015) *Arbiom*, USA. Online: http://www.arbiom.com/arbiom-announces-bioskog-a-biorefinery-project-in-golbey-france-in-partnership-with-norske-skog-golbey/ (assessed 16th May 2018).
21. News: DuPont Celebrates the Opening of the World's Largest Cellu-losic Ethanol Plant (October 2015) *DuPont*, USA. Online: http://www.dupont.com/corporate-functions/media-cen-ter/press-releases/dupont-celebrates-opening-of-worlds-largest-cellulosic-ethanol-plant.html (assessed 16th May 2018).
22. News: DuPont to Sell Cellulosic Ethanol Plant in Blow to Biofuel (November 2017) *Reuters*, UK. Online: https://in.reuters.com/arti-cle/us-dowdupont-ethanol/dupont-to-sell-cellulosic-ethanol-plant-in-blow-to-biofuel-idINKBN1D22T5 (assessed 16th May 2018).
23. News: Hugoton Cellulosic Ethanol Plant Sold Out of Bankruptcy (De-cember 2016) *HPJ*, USA. Online: http://www.hpj.com/ag_news/hugoton-cellulosic-ethanol-plant-sold-out-of-bankruptcy/article_ae8fb952-c85f-11e6-87dc-0b12cf1982e3.html (assessed 16th May 2018).
24. Why Hasn't Cellulosic Ethanol Taken Over, Like it Was Supposed To? (November 2013) *The Washington Post*, USA. Online: https://www.washingtonpost.com/business/economy/why-hasnt-cellulosic-ethanol-taken-over-like-it-was-supposed-to/2013/11/08/b25b0d2c-466a-11e3-a196-3544a03c2351_story.html?noredi-rect=on&utm_term=.a18e1765826f (assessed 16th May 2018).
25. POET-DSM Advanced Biofuels (2014) *POET-DSM*, USA. Online: http://poet-dsm.com/ (assessed 16th May 2018).
26. From Waste to Cellulosic Ethanol, Biomethanol (2018) *Enerkem*, Canada. Online: https://enerkem.com/ (assessed 16th May 2018).
27. St1 Biofuels (2018) *St1*, Finland. Online: http://www.st1biofu-els.com/ (assessed 16th May 2018).
28. Kobayashi, H., Yabushita, M., Komanoya, T., Hara, K., Fujita, I., and Fukuoka, A. (2013) High-yielding one-pot synthesis of glucose from cellulose using simple activated carbons and trace hydrochloric acid. ACS Catalysis, **3**(4), 581-587.

Index